Biomolecular Feedback Systems

Biomolecular Feedback Systems

Domitilla Del Vecchio
Richard M. Murray

Princeton University Press
Princeton and Oxford

Published by Princeton University Press, 41 William Street, Princeton, New Jersey 08540
In the United Kingdom: Princeton University Press, 6 Oxford Street, Woodstock, Oxfordshire OX20 1TW

press.princeton.edu

Library of Congress Cataloging-in-Publication Data

Del Vecchio, Domitilla, 1975–
    Biomolecular feedback systems / Domitilla Del Vecchio and Richard M. Murray.
        pages cm
    Includes bibliographical references and index.
    ISBN 978-0-691-16153-2 (hardcover : alk. paper) 1. Biological Control Systems. 2. Systems biology. I. Murray, Richard M. II. Title.
    QH508.D45 2015
    612.8′233–dc23   2014013061

British Library Cataloging-in-Publication Data is available

The publisher would like to acknowledge the authors of this volume for providing the camera-ready copy from which this book was printed.

This book has been composed in LaTeX

Printed on acid-free paper. ∞

Printed in the United States of America

10 9 8 7 6 5 4 3 2 1

Contents

Preface

This text is intended for researchers interested in the application of feedback and control to biomolecular systems. The material has been designed so that it can be used in parallel with the textbook *Feedback Systems* [1] as part of a course on biomolecular feedback and control systems, or as a stand-alone reference for readers who have had a basic course in feedback and control theory. The full text for this book, along with additional supplemental material, is available on a companion Web site:

http://press.princeton.edu/titles/10285.html

The material in this book is intended to be useful to three overlapping audiences: graduate students in biology and bioengineering interested in understanding the role of feedback in natural and engineered biomolecular systems; advanced undergraduates and graduate students in engineering disciplines who are interested in the use of feedback in biological circuit design; and established researchers in the biological sciences who want to explore the potential application of principles and tools from control theory to biomolecular systems. We have written the text assuming some familiarity with basic concepts in feedback and control, but have tried to provide insights and specific results as needed, so that the material can be learned in parallel. We also assume some familiarity with cell biology, at the level of a first course for non-majors. The individual chapters in the text indicate the prerequisites in more detail, most of which are covered either in Åström and Murray [1] or in the supplemental information available from the companion Web site.

Acknowledgments. Many colleagues and students provided feedback and advice on the book. We would particularly like to thank Mustafa Khammash, Eric Klavins, and Eduardo Sontag, who gave detailed comments on some of the early versions of the text. In addition, we would like to acknowledge Abdullah Amadeh, Andras Gyorgy, Narmada Herath, Yutaka Hori, Shridhar Jayanthi, Scott Livingston, Rob Phillips, Yili Qian, Phillip Rivera, Vipul Singhal, Anandh Swaminathan, Eric Winfree, and Enoch Yeung for their support and comments along the way. Finally, we would like to thank Caltech, MIT and the University of Michigan for providing the many resources that were necessary to bring this book to fruition.

Domitilla Del Vecchio   Richard M. Murray
Cambridge, Massachusetts   Pasadena, California

Biomolecular Feedback Systems

Chapter 1
Introductory Concepts

This chapter provides a brief introduction to concepts from systems biology, tools from differential equations and control theory, and approaches to modeling, analysis and design of biomolecular feedback systems. We begin with a discussion of the role of modeling, analysis and feedback in biological systems. This is followed by a short review of key concepts and tools from control and dynamical systems theory, intended to provide insight into the main methodology described in the text. Finally, we give a brief introduction to the field of synthetic biology, which is the primary topic of the latter portion of the text. Readers who are familiar with one or more of these areas can skip the corresponding sections without loss of continuity.

1.1 Systems biology: Modeling, analysis and role of feedback

At a variety of levels of organization—from molecular to cellular to organismal—biology is becoming more accessible to approaches that are commonly used in engineering: mathematical modeling, systems theory, computation and abstract approaches to synthesis. Conversely, the accelerating pace of discovery in biological science is suggesting new design principles that may have important practical applications in human-made systems. This synergy at the interface of biology and engineering offers many opportunities to meet challenges in both areas. The guiding principles of feedback and control are central to many of the key questions in biological science and engineering and can play an enabling role in understanding the complexity of biological systems.

In this section we summarize our view on the role that modeling and analysis should (eventually) play in the study of biological systems, and discuss some of the ways in which an understanding of feedback principles in biology can help us better understand and design complex biomolecular circuits.

There are a wide variety of biological phenomena that provide a rich source of examples for control, including gene regulation and signal transduction; hormonal, immunological, and cardiovascular feedback mechanisms; muscular control and locomotion; active sensing, vision, and proprioception; attention and consciousness; and population dynamics and epidemics. Each of these (and many more) provide opportunities to figure out what works, how it works, and what can be done to affect it. Our focus here is at the molecular scale, but the principles and approach that we describe can also be applied at larger time and length scales.

Modeling and analysis

Over the past several decades, there have been significant advances in modeling capabilities for biological systems that have provided new insights into the complex interactions of the molecular-scale processes that implement life. Reduced-order modeling has become commonplace as a mechanism for describing and documenting experimental results, and high-dimensional stochastic models can now be simulated in reasonable periods of time to explore underlying stochastic effects. Coupled with our ability to collect large amounts of data from flow cytometry, micro-array analysis, single-cell microscopy, and other modern experimental techniques, our understanding of biomolecular processes is advancing at a rapid pace.

Unfortunately, although models are becoming much more common in biological studies, they are still far from playing the central role in explaining complex biological phenomena. Although there are exceptions, the predominant use of models is to "document" experimental results: a hypothesis is proposed and tested using careful experiments, and then a model is developed to match the experimental results and help demonstrate that the proposed mechanisms can lead to the observed behavior. This necessarily limits our ability to explain complex phenomena to those for which controlled experimental evidence of the desired phenomena can be obtained.

This situation is much different than standard practice in the physical sciences and engineering, as illustrated in Figure 1.1 (in the context of modeling, analysis, and control design for gas turbine aeroengines). In those disciplines, experiments are routinely used to help build models for individual components at a variety of levels of detail, and then these component-level models are interconnected to obtain a system-level model. This system-level model, carefully built to capture the appropriate level of detail for a given question or hypothesis, is used to explain, predict, and systematically analyze the behaviors of a system. Because of the ways in which models are viewed, it becomes possible to prove (or invalidate) a hypothesis through analysis of the model, and the fidelity of the models is such that decisions can be made based on them. Indeed, in many areas of modern engineering— including electronics, aeronautics, robotics, and chemical processing, to name a few—models play a primary role in the understanding of the underlying physics and/or chemistry, and these models are used in predictive ways to explore design tradeoffs and failure scenarios.

A key element in the successful application of modeling in engineering disciplines is the use of *reduced-order models* that capture the underlying dynamics of the system without necessarily modeling every detail of the underlying mechanisms. These reduced-order models are often coupled with schematics diagrams, such as those shown in Figure 1.2, to provide a high level view of a complex system. The generation of these reduced-order models, either directly from data or through analytical or computational methods, is critical in the effective applica-

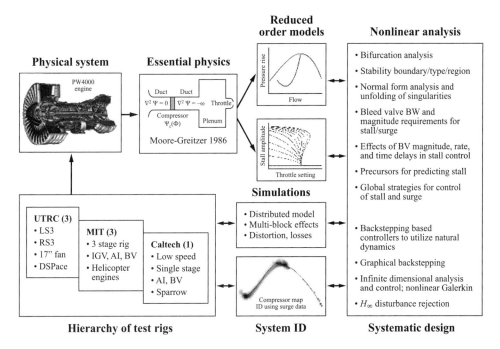

Figure 1.1: Sample modeling, analysis and design framework for an engineering system. The physical system (in this case a jet engine) is first modeled using a detailed mathematical description that captures the essential physics that are relevant for the design. Reduced-order models (typically differential equations and steady state input/output maps) are then created for use in analysis and design. A list of some typical tools in this domain are shown in the box on the right. These tools are used to design the system and then combined with simulations and system identification tools. Finally, a hierarchy of testing environments are used as the system is built and tested, finally resulting in an implementation of the full system. Additional details can be found in [29].

tion of modeling since modeling of the detailed mechanisms produces high fidelity models that are too complicated to use with existing tools for analysis and design. One area in which the development of reduced-order models is fairly advanced is in control theory, where input/output models, such as block diagrams and transfer functions, are used to capture structured representations of dynamics at the appropriate level of fidelity for the task at hand [1].

While developing predictive models and corresponding analysis tools for biology is much more difficult, it is perhaps even more important that biology make use of models, particularly reduced-order models, as a central element of understanding. Biological systems are by their nature extremely complex and can behave in counterintuitive ways. Only by capturing the many interacting aspects of the system in a formal model can we ensure that we are reasoning properly about its behavior, especially in the presence of uncertainty. To do this will require substantial effort in building models that capture the relevant dynamics at the proper

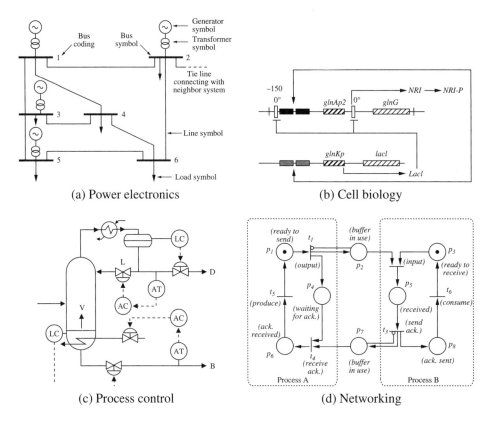

Figure 1.2: Schematic diagrams representing models in different disciplines. Each diagram is used to illustrate the dynamics of a feedback system: (a) electrical schematics for a power system [58], (b) a biological circuit diagram for a synthetic clock circuit [6], (c) a process diagram for a distillation column [86] and (d) a Petri net description of a communication protocol.

scales (depending on the question being asked), as well as building an analytical framework for answering questions of biological relevance.

The good news is that a variety of new techniques, ranging from experiments to computation to theory, are enabling us to explore new approaches to modeling that attempt to address some of these challenges. In this text we focus on the use of relevant classes of reduced-order models that can be used to capture many phenomena of biological relevance.

Dynamic behavior and phenotype

One of the key needs in developing a more systematic approach to the use of models in biology is to become more rigorous about the various behaviors that are important for biological systems. One of the key concepts that needs to be formalized

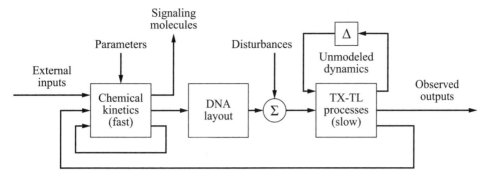

Figure 1.3: Conceptual modeling framework for biomolecular feedback systems. The chemical kinetics block represents reactions between molecular species, resulting in signaling molecules and bound promoters. The DNA layout block accounts for the organization of the DNA, which may be "rewired" to achieve a desired function. The TX-TL processes block represents the core transcription and translation processes, which are often much slower than the reactions between various species. The inputs and outputs of the various blocks represent interconnections and external interactions.

is the notion of "phenotype." This term is often associated with the existence of an equilibrium point in a reduced-order model for a system, but clearly more complex (non-equilibrium) behaviors can occur and the "phenotypic response" of a system to an input may not be well-modeled by a steady operating condition. Even more problematic is determining which regulatory structures are "active" in a given phenotype (versus those for which there is a regulatory pathway that is saturated and hence not active).

Figure 1.3 shows a graphical representation of a class of systems that captures many of the features we are interested in. The chemical kinetics of the system are typically modeled using mass action kinetics (reaction rate equations) and represent the fast dynamics of chemical reactions. The reactions include the binding of activators and repressors to DNA, as well as the initiation of transcription. The DNA layout block represents the physical layout of the DNA, which determines which genes are controlled by which promoters. The core processes of transcription (TX) and translation (TL) represent the slow dynamics (relative to the chemical kinetics) of protein expression (including maturation).

Several other inputs and outputs are represented in the figure. In the chemical kinetics block, we allow external inputs, such as chemical inducers, and external parameters (rate parameters, enzyme concentrations, etc.) that will affect the reactions that we are trying to capture in our model. We also include a (simplified) notion of disturbances, represented in the diagram as an external input that affects the rate of transcription. This disturbance is typically a stochastic input that represents the fact that gene expression can be noisy. In terms of outputs, we capture two possibilities in the diagram: small molecule outputs—often used for signaling to other subsystems but which could include outputs from metabolic processes—

and protein outputs, such as as fluorescent reporters.

Another feature of the diagram is the block labeled "unmodeled dynamics," which represents the fact that our models of the core processes of gene expression are likely to be simplified models that ignore many details. These dynamics are modeled as a feedback interconnection with transcription and translation, which turns out to provide a rich framework for application of tools from control theory (but unfortunately one that we will not explore in great detail within this text). Tools for understanding this class of uncertainty are available for both linear and nonlinear control systems [1] and allow stability and performance analyses in the presence of uncertainty.

The combination of partially unknown parameters, external disturbances, and unmodeled dynamics are collectively referred to as *model uncertainty* and are an important element of our analysis of biomolecular feedback systems. Often we will analyze the behavior of a system assuming that the parameters are known, disturbances are small and our models are accurate. This analysis can give valuable insights into the behavior of the system, but it is important to verify that this behavior is robust with respect to uncertainty, a topic that we will discuss in Chapter 3.

A somewhat common situation is that a system may have multiple equilibrium points and the "phenotype" of the system is represented by the particular equilibrium point that the system converges to. In the simplest case, we can have *bistability*, in which there are two equilibrium points for a fixed set of parameters. Depending on the initial conditions and external inputs, a given system may end up near one equilibrium point or the other, providing two distinct phenotypes. A model with bistability (or multi-stability) provides one method of modeling memory in a system: the cell or organism remembers its history by virtue of the equilibrium point to which it has converted.

For more complex phenotypes, where the subsystems are not at a steady operating point, one can consider temporal patterns such as limit cycles (periodic orbits) or non-equilibrium input/output responses. Analysis of these more complicated behaviors requires more sophisticated tools, but again model-based analysis of stability and input/output responses can be used to characterize the phenotypic behavior of a biological system under different conditions or contexts.

Additional types of analysis that can be applied to systems of this form include sensitivity analysis (dependence of solution properties on selected parameters), uncertainty analysis (impact of disturbances, unknown parameters and unmodeled dynamics), bifurcation analysis (changes in phenotype as a function of input levels, context or parameters) and probabilistic analysis (distributions of states as a function of distributions of parameters, initial conditions or inputs). In each of these cases, there is a need to extend existing tools to exploit the particular structure of the problems we consider, as well as modify the techniques to provide relevance to biological questions.

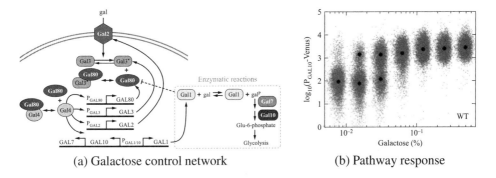

(a) Galactose control network   (b) Pathway response

Figure 1.4: Galactose response in yeast [96]. (a) GAL signaling circuitry showing a number of different feedback pathways that are used to detect the presence of galactose and switch on the metabolic pathway. (b) Pathway activity as a function of galactose concentration. The points at each galactose concentration represent the activity level of the galactose metabolic pathway in an individual cell. Black dots indicate the mean of a Gaussian mixture model classification [96]. Small random deviations were added to each galactose concentration (horizontal axis) to better visualize the distributions. Figures adapted from [96].

Stochastic behavior

Another important feature of many biological systems is stochasticity: biological responses have an element of randomness so that even under carefully controlled conditions, the response of a system to a given input may vary from experiment to experiment. This randomness can have many possible sources, including external perturbations that are modeled as stochastic processes and internal processes such as molecular binding and unbinding, whose stochasticity stems from the underlying thermodynamics of molecular reactions.

While for many engineered systems it is common to try to eliminate stochastic behavior (yielding a "deterministic" response), for biological systems there appear to be many situations in which stochasticity is important for the way in which organisms survive. In biology, nothing is 100% and so there is always some chance that two identical organisms will respond differently. Thus viruses are never completely contagious and so some organisms will survive, and DNA replication is never error free, and so mutations and evolution can occur. In studying circuits where these types of effects are present, it thus becomes important to study the distribution of responses of a given biomolecular circuit, and to collect data in a manner that allows us to quantify these distributions.

One important indication of stochastic behavior is *bimodality*. We say that a circuit or system is bimodal if the response of the system to a given input or condition has two or more distinguishable classes of behaviors. An example of bimodality is shown in Figure 1.4, which shows the response of the galactose metabolic machinery in yeast. We see from the figure that even though genetically identical organisms are exposed to the same external environment (a fixed galactose concentration), the amount of activity in individual cells can have a large amount of variability. At some concentrations there are clearly two subpopulations of cells: those in which the galactose metabolic pathway is turned on (higher reporter fluorescence values on the y axis) and those for which it is off (lower reporter fluorescence).

Another characterization of stochasticity in cells is the separation of noisiness in protein expression into two categories: "intrinsic" noise and "extrinsic" noise. Roughly speaking, extrinsic noise represents variability in gene expression that affects all proteins in the cell in a correlated way. Extrinsic noise can be due to environmental changes that affect the entire cell (temperature, pH, oxygen level) or global changes in internal factors such as energy or metabolite levels (perhaps due to metabolic loading). Intrinsic noise, on the other hand, is the variability due to the inherent randomness of molecular events inside the cell and represents a collection of independent random processes. One way to attempt to measure the amount of intrinsic and extrinsic noise is to take two identical copies of a biomolecular circuit and compare their responses [27, 92]. Correlated variations in the output of the circuits corresponds (roughly) to extrinsic noise and uncorrelated variations to intrinsic noise [43, 92].

The types of models that are used to capture stochastic behavior are very different than those used for deterministic responses. Instead of writing differential equations that track average concentration levels, we must keep track of the individual events that can occur with some probability per unit time (or "propensity"). We will explore the methods for modeling and analysis of stochastic systems in Chapter 4.

1.2   The cell as a system

The molecular processes inside a cell determine its behavior and are responsible for metabolizing nutrients, generating motion, enabling procreation and carrying out the other functions of the organism. In multi-cellular organisms, different types of cells work together to enable more complex functions. In this section we briefly describe the role of dynamics and control within a cell and discuss the basic processes that govern its behavior and its interactions with its environment. We assume knowledge of the basics of cell biology at the level found in standard textbooks on cell biology such as Alberts et al. [2] or Phillips et al. [78].

Figure 1.5 shows a schematic of the major components in the cell: sensing, signaling, regulation, and metabolism. Sensing of environmental signals typically occurs through membrane receptors that are specific to different molecules. Cells can also respond to light or pressure, allowing the cell to sense the environment, including other cells. There are several types of receptors, some allow the signaling molecules in the environment to enter the cell wall, such as in the case of ion

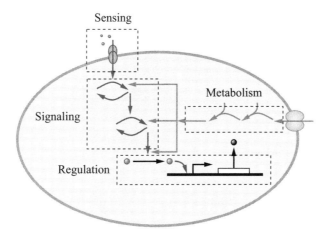

Figure 1.5: The cell as a system. The major subsystems are sensing, signaling, regulation, and metabolism.

channels. Others activate proteins on the internal part of the cell membrane once they externally bind to the signaling molecule, such as enzyme-linked receptors or G-protein coupled receptors.

As a consequence of the sensing, a cascade of signal transduction occurs (signaling) in which proteins are sequentially activated by (usually) receiving phosphate groups from ATP molecules through the processes of phosphorylation and/or phosphotransfer. These cascades transmit information to downstream processes, such as gene expression, by amplifying the information and dynamically filtering signals to select for useful features. The temporal dynamics of environmental signals and the kinetic properties of the stages in the signaling cascades determine how a signal is transmitted/filtered. At the bottom stages of signaling cascades, proteins are activated to become transcription factors, which can activate or repress the expression of other proteins through regulation of gene expression. The temporal dynamics of this regulation, with time scales in the range of minutes to hours, are usually much slower than that of the transmission in the signaling pathway, which has time scales ranging from subseconds to seconds. Metabolic pathways, such as the glycolysis pathway, also characterized by very fast time scales, are in charge of producing the necessary resources for all the other processes in the cells. Through these pathways, nutrients in the environment, such as glucose, are broken down through a series of enzymatic reactions, producing, among other products, ATP, which is the energy currency in the cell used for many of the reactions, including those involved in signaling and gene expression.

Example: Chemotaxis

As an example of a sensing-transmission-actuation process in the cell, we consider *chemotaxis*, the process by which microorganisms move in response to chemical

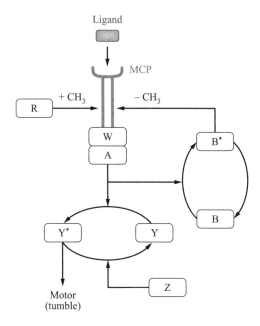

Figure 1.6: A simplified circuit diagram for chemotaxis, showing the biomolecular processes involved in regulating flagellar motion. Figure adapted from Rao et al. [83].

stimuli. Examples of chemotaxis include the ability of organisms to move in the direction of nutrients or move away from toxins in the environment. Chemotaxis is called *positive chemotaxis* if the motion is in the direction of the stimulus and *negative chemotaxis* if the motion is away from the stimulus.

The chemotaxis system in *E. coli* consists of a sensing system that detects the presence of nutrients, an actuation system that propels the organism in its environment, and control circuitry that determines how the cell should move in the presence of chemicals that stimulate the sensing system. The main components of the control circuitry are shown in Figure 1.6. The sensing component is responsible for detecting the presence of ligands in the environment and initiating signaling cascades. The computation component, realized through a combination of protein phosphorylation and methylation, implements a feedback (integral) controller that allows the bacterium to adapt to changes in the environmental ligand concentration. This adaptation occurs by an actuator that allows the bacterium to ultimately move in the direction in which the ligand concentration increases.

The actuation system in the *E. coli* consists of a set of flagella that can be spun using a flagellar motor embedded in the outer membrane of the cell, as shown in Figure 1.7a. When the flagella all spin in the counterclockwise direction, the individual flagella form a bundle and cause the organism to move roughly in a straight line. This behavior is called a "run" motion. Alternatively, if the flagella spin in the clockwise direction, the individual flagella do not form a bundle and the organ-

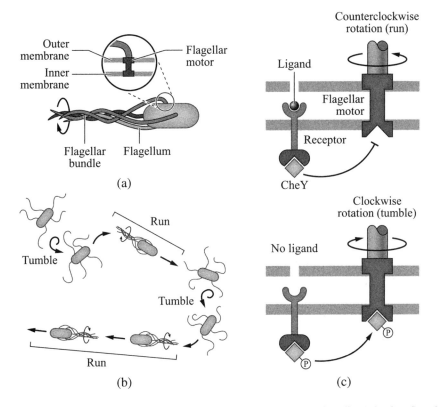

Figure 1.7: Bacterial chemotaxis. (a) Flagellar motors are responsible for spinning flagella. (b) When flagella spin in the clockwise direction, the organism tumbles, while when they spin in the counterclockwise direction, the organism runs. (c) The direction in which the flagella spin is determined by whether the CheY protein is phosphorylated. Figures adapted from Phillips, Kondev and Theriot [78].

ism "tumbles," causing it to rotate (Figure 1.7b). The selection of the motor direction is controlled by the protein CheY: if phosphorylated CheY binds to the motor complex, the motor spins clockwise (tumble), otherwise it spins counterclockwise (run). As a consequence, the chemotaxis mechanism is stochastic in nature, with biased random motions causing the average behavior to be either positive, negative, or neutral (in the absence of stimuli).

1.3 Control and dynamical systems tools[1]

To study the complex dynamics and feedback present in biological systems, we will make use of mathematical models combined with analytical and computational tools. In this section we present a brief introduction to some of the key concepts

---

[1]The material in this section is adapted from *Feedback Systems*, Chapter 1 [1].

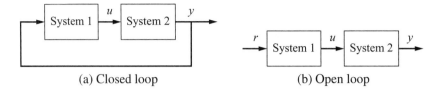

(a) Closed loop   (b) Open loop

Figure 1.8: Open and closed loop systems. (a) The output of system 1 is used as the input of system 2, and the output of system 2 becomes the input of system 1, creating a closed loop system. (b) The interconnection between system 2 and system 1 is removed, and the system is said to be open loop.

from control and dynamical systems that are relevant for the study of biomolecular systems considered in later chapters. More details on the application of specific concepts listed here to biomolecular systems is provided in the main body of the text. Readers who are familiar with introductory concepts in dynamical systems and control, at the level described in Åström and Murray [1], for example, can skip this section.

Dynamics, feedback and control

A *dynamical system* is a system whose behavior changes over time, often in response to external stimulation or forcing. The term *feedback* refers to a situation in which two (or more) dynamical systems are connected together such that each system influences the other and their dynamics are thus strongly coupled. Simple causal reasoning about a feedback system is difficult because the first system influences the second and the second system influences the first, leading to a circular argument. This makes reasoning based on cause and effect tricky, and it is necessary to analyze the system as a whole. A consequence of this is that the behavior of feedback systems is often counterintuitive, and it is therefore often necessary to resort to formal methods to understand them.

Figure 1.8 illustrates in block diagram form the idea of feedback. We often use the terms *open loop* and *closed loop* when referring to such systems. A system is said to be a closed loop system if the systems are interconnected in a cycle, as shown in Figure 1.8a. If we break the interconnection, we refer to the configuration as an open loop system, as shown in Figure 1.8b.

Biological systems make use of feedback in an extraordinary number of ways, on scales ranging from molecules to cells to organisms to ecosystems. One example is the regulation of glucose in the bloodstream through the production of insulin and glucagon by the pancreas. The body attempts to maintain a constant concentration of glucose, which is used by the body's cells to produce energy. When glucose levels rise (after eating a meal, for example), the hormone insulin is released and causes the body to store excess glucose in the liver. When glucose levels are low, the pancreas secretes the hormone glucagon, which has the opposite effect. Referring to Figure 1.8, we can view the liver as system 1 and the pancreas as system 2. The output from the liver is the glucose concentration in the blood, and the output from the pancreas is the amount of insulin or glucagon produced. The interplay between insulin and glucagon secretions throughout the day helps to keep the blood-glucose concentration constant, at about 90 mg per 100 mL of blood.

Feedback has many interesting properties that can be exploited in designing systems. As in the case of glucose regulation, feedback can make a system resilient toward external influences. It can also be used to create linear behavior out of nonlinear components, a common approach in electronics. More generally, feedback allows a system to be insensitive both to external disturbances and to variations in its individual elements.

Feedback has potential disadvantages as well. It can create dynamic instabilities in a system, causing oscillations or even runaway behavior. Another drawback, especially in engineering systems, is that feedback can introduce unwanted sensor noise into the system, requiring careful filtering of signals. It is for these reasons that a substantial portion of the study of feedback systems is devoted to developing an understanding of dynamics and a mastery of techniques in dynamical systems.

Feedback properties

Feedback is a powerful idea that is used extensively in natural and technological systems. The principle of feedback is simple: implement correcting actions based on the difference between desired and actual performance. In engineering, feedback has been rediscovered and patented many times in many different contexts. The use of feedback has often resulted in vast improvements in system capability, and these improvements have sometimes been revolutionary in areas such as power generation and transmission, aerospace and transportation, materials and processing, instrumentation, robotics and intelligent machines, and networking and communications. The reason for this is that feedback has some truly remarkable properties, which we discuss briefly here.

Robustness to uncertainty. One of the key uses of feedback is to provide robustness to uncertainty. By measuring the difference between the sensed value of a regulated signal and its desired value, we can supply a corrective action. If the system undergoes some change that affects the regulated signal, then we sense this change and try to force the system back to the desired operating point.

As an example of this principle, consider the simple feedback system shown in Figure 1.9. In this system, the speed of a vehicle is controlled by adjusting the amount of gas flowing to the engine. Simple *proportional-integral* (PI) feedback is used to make the amount of gas depend on both the error between the current and the desired speed and the integral of that error. The plot in Figure 1.9b shows the results of this feedback for a step change in the desired speed and a variety of

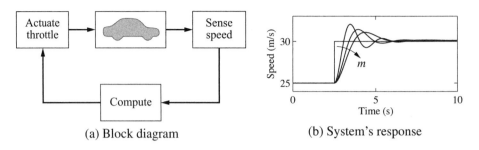

(a) Block diagram   (b) System's response

Figure 1.9: A feedback system for controlling the speed of a vehicle. (a) In the block diagram, the speed of the vehicle is measured and compared to the desired speed within the "Compute" block. Based on the difference in the actual and desired speeds, the throttle (or brake) is used to modify the force applied to the vehicle by the engine, drivetrain and wheels. (b) The figure shows the response of the control system to a commanded change in speed from 25 m/s to 30 m/s. The three different curves correspond to differing masses of the vehicle, between 1000 and 3000 kg, demonstrating the robustness of the closed loop system to a very large change in the vehicle characteristics.

different masses for the car, which might result from having a different number of passengers or towing a trailer. Notice that independent of the mass (which varies by a factor of 3!), the steady-state speed of the vehicle always approaches the desired speed and achieves that speed within approximately 5 s. Thus the performance of the system is robust with respect to this uncertainty.

Another early example of the use of feedback to provide robustness is the negative feedback amplifier. When telephone communications were developed, amplifiers were used to compensate for signal attenuation in long lines. A vacuum tube was a component that could be used to build amplifiers. Distortion caused by the nonlinear characteristics of the tube amplifier together with amplifier drift were obstacles that prevented the development of line amplifiers for a long time. A major breakthrough was the invention of the feedback amplifier in 1927 by Harold S. Black, an electrical engineer at Bell Telephone Laboratories. Black used *negative feedback*, which reduces the gain but makes the amplifier insensitive to variations in tube characteristics. This invention made it possible to build stable amplifiers with linear characteristics despite the nonlinearities of the vacuum tube amplifier.

Feedback is also pervasive in biological systems, where transcriptional, translational and allosteric mechanisms are used to regulate internal concentrations of various species, and much more complex feedbacks are used to regulate properties at the organism level (such as body temperature, blood pressure and circadian rhythm). One difference in biological systems is that the separation of sensing, actuation and computation, a common approach in most engineering control systems, is less evident. Instead, the dynamics of the molecules that sense the environmental condition and make changes to the operation of internal components may be integrated together in ways that make it difficult to untangle the operation of the system. Similarly, the "reference value" to which we wish to regulate a system may not be an explicit signal, but rather a consequence of many different changes in the dynamics that are coupled back to the regulatory elements. Hence we do not see a clear "set point" for the desired ATP concentration, blood oxygen level or body temperature, for example. These difficulties complicate our analysis of biological systems, though many important insights can still be obtained.

Design of dynamics. Another use of feedback is to change the dynamics of a system. Through feedback, we can alter the behavior of a system to meet the needs of an application: systems that are unstable can be stabilized, systems that are sluggish can be made responsive and systems that have drifting operating points can be held constant. Control theory provides a rich collection of techniques to analyze the stability and dynamic response of complex systems and to place bounds on the behavior of such systems by analyzing the gains of linear and nonlinear operators that describe their components.

An example of the use of control in the design of dynamics comes from the area of flight control. The following quote, from a lecture presented by Wilbur Wright to the Western Society of Engineers in 1901 [70], illustrates the role of control in the development of the airplane:

> Men already know how to construct wings or airplanes, which when driven through the air at sufficient speed, will not only sustain the weight of the wings themselves, but also that of the engine, and of the engineer as well. Men also know how to build engines and screws of sufficient lightness and power to drive these planes at sustaining speed . . . Inability to balance and steer still confronts students of the flying problem . . . When this one feature has been worked out, the age of flying will have arrived, for all other difficulties are of minor importance.

The Wright brothers thus realized that control was a key issue to enable flight. They resolved the compromise between stability and maneuverability by building an airplane, the Wright Flyer, that was unstable but maneuverable. The Flyer had a rudder in the front of the airplane, which made the plane very maneuverable. A disadvantage was the necessity for the pilot to keep adjusting the rudder to fly the plane: if the pilot let go of the stick, the plane would crash. Other early aviators tried to build stable airplanes. These would have been easier to fly, but because of their poor maneuverability they could not be brought up into the air. By using their insight and skillful experiments, the Wright brothers made the first successful flight at Kitty Hawk in 1903.

Since it was quite tiresome to fly an unstable aircraft, there was strong motivation to find a mechanism that would stabilize an aircraft. Such a device, invented by Sperry, was based on the concept of feedback. Sperry used a gyro-stabilized pendulum to provide an indication of the vertical. He then arranged a feedback

Figure 1.10: Aircraft autopilot system. The Sperry autopilot (left) contained a set of four gyros coupled to a set of air valves that controlled the wing surfaces. The 1912 Curtiss used an autopilot to stabilize the roll, pitch and yaw of the aircraft and was able to maintain level flight as a mechanic walked on the wing (right) [45].

mechanism that would pull the stick to make the plane go up if it was pointing down, and vice versa. The Sperry autopilot was the first use of feedback in aeronautical engineering, and Sperry won a prize in a competition for the safest airplane in Paris in 1914. Figure 1.10 shows the Curtiss seaplane and the Sperry autopilot. The autopilot is a good example of how feedback can be used to stabilize an unstable system and hence "design the dynamics" of the aircraft.

One of the other advantages of designing the dynamics of a device is that it allows for increased modularity in the overall system design. By using feedback to create a system whose response matches a desired profile, we can hide the complexity and variability that may be present inside a subsystem. This allows us to create more complex systems by not having to simultaneously tune the responses of a large number of interacting components. This was one of the advantages of Black's use of negative feedback in vacuum tube amplifiers: the resulting device had a well-defined linear input/output response that did not depend on the individual characteristics of the vacuum tubes being used.

Drawbacks of feedback. While feedback has many advantages, it also has some drawbacks. Chief among these is the possibility of instability if the system is not designed properly. We are all familiar with the undesirable effects of feedback when the amplification on a microphone is turned up too high in a room. This is an example of feedback instability, something that we obviously want to avoid. This is tricky because we must design the system not only to be stable under nominal conditions but also to remain stable under all possible perturbations of the dynamics. In biomolecular systems, these types of instabilities may exhibit themselves as situations in which cells no longer function properly due to over expression of engineered genetic components, or small fluctuations in parameters may cause the system to suddenly cease to function properly.

In addition to the potential for instability, feedback inherently couples different parts of a system. One common problem is that feedback often injects "crosstalk" into the system. By coupling different parts of a biomolecular circuit, the fluctuations in one part of the circuit affect other parts, which themselves may couple to the initial source of the fluctuations. If we are designing a biomolecular system, this crosstalk may affect our ability to design independent "modules" whose behavior can be described in isolation.

Coupled to the problem of crosstalk is the substantial increase in complexity that results when embedding multiple feedback loops in a system. An early engineering example of this was the use of microprocessor-based feedback systems in automobiles. The use of microprocessors in automotive applications began in the early 1970s and was driven by increasingly strict emissions standards, which could be met only through electronic controls. Early systems were expensive and failed more often than desired, leading to frequent customer dissatisfaction. It was only through aggressive improvements in technology that the performance, reliability and cost of these systems allowed them to be used in a transparent fashion. Even today, the complexity of these systems is such that it is difficult for an individual car owner to fix problems. While nature has evolved many feedback structures that are robust and reliable, engineered biomolecular systems are still quite rudimentary and we can anticipate that as we increase the use of feedback to compensate for uncertainty, we will see a similar period in which engineers must overcome a steep learning curve before we can get robust and reliable behavior as a matter of course.

Feedforward. Feedback is reactive: there must be an error before corrective actions are taken. However, in some circumstances it is possible to measure a disturbance before it enters the system, and this information can then be used to take corrective action before the disturbance has influenced the system. The effect of the disturbance is thus reduced by measuring it and generating a control signal that counteracts it. This way of controlling a system is called *feedforward*. Feedforward is particularly useful in shaping the response to command signals because command signals are always available. Since feedforward attempts to match two signals, it requires good process models; otherwise the corrections may have the wrong size or may be badly timed.

The ideas of feedback and feedforward are very general and appear in many different fields. In economics, feedback and feedforward are analogous to a market-based economy versus a planned economy. In business, a feedforward strategy corresponds to running a company based on extensive strategic planning, while a feedback strategy corresponds to a reactive approach. In biology, feedforward has been suggested as an essential element for motion control in humans that is tuned during training. Experience indicates that it is often advantageous to combine feedback and feedforward, and the correct balance requires insight and understanding of their respective properties.

Positive feedback. In most of control theory, the emphasis is on the role of *negative feedback*, in which we attempt to regulate the system by reacting to disturbances in a way that decreases the effect of those disturbances. In some systems, particularly biological systems, *positive feedback* can play an important role. In a system with positive feedback, the increase in some variable or signal leads to a situation in which that quantity is further increased through its dynamics. This has a destabilizing effect and is usually accompanied by a saturation that limits the growth of the quantity. Although often considered undesirable, this behavior is used in biological (and engineering) systems to obtain a very fast response to a condition or signal.

One example of the use of positive feedback is to create switching behavior, in which a system maintains a given state until some input crosses a threshold. Hysteresis is often present so that noisy inputs near the threshold do not cause the system to jitter. This type of behavior is called *bistability* and is often associated with memory devices.

1.4   Input/output modeling[2]

A model is a mathematical representation of a physical, biological or information system. Models allow us to reason about a system and make predictions about how a system will behave. In this text, we will mainly be interested in models of dynamical systems describing the input/output behavior of systems, and we will often work in "state space" form. In the remainder of this section we provide an overview of some of the key concepts in input/output modeling. The mathematical details introduced here are explored more fully in Chapter 3.

The heritage of electrical engineering

The approach to modeling that we take builds on the view of models that emerged from electrical engineering, where the design of electronic led to a focus on input/output behavior. A system was considered a device that transforms inputs to outputs, as illustrated in Figure 1.11. Conceptually an input/output model can be viewed as a giant table of inputs and outputs. Given an input signal $u(t)$ over some interval of time, the model should produce the resulting output $y(t)$.

The input/output framework is used in many engineering disciplines since it allows us to decompose a system into individual components connected through their inputs and outputs. Thus, we can take a complicated system such as a radio or a television and break it down into manageable pieces such as the receiver, demodulator, amplifier and speakers. Each of these pieces has a set of inputs and outputs

---

[2]The material in this section is adapted from *Feedback Systems*, Sections 2.1–2.2 [1].

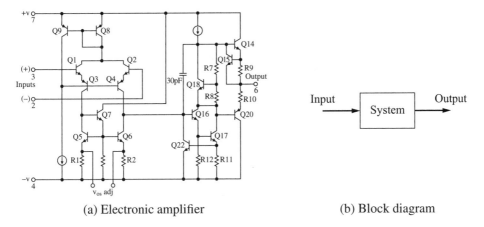

(a) Electronic amplifier   (b) Block diagram

Figure 1.11: Illustration of the input/output view of a dynamical system. (a) The figure shows a detailed circuit diagram for an electronic amplifier; the one in (b) is its representation as a block diagram.

and, through proper design, these components can be interconnected to form the entire system.

The input/output view is particularly useful for the special class of *linear time-invariant systems*. This term will be defined more carefully below, but roughly speaking a system is linear if the superposition (addition) of two inputs yields an output that is the sum of the outputs that would correspond to individual inputs being applied separately. A system is time-invariant if the output response for a given input does not depend on when that input is applied. While most biomolecular systems are neither linear nor time-invariant, they can often be approximated by such models, often by looking at perturbations of the system from its nominal behavior, in a fixed context.

One of the reasons that linear time-invariant systems are so prevalent in modeling of input/output systems is that a large number of tools have been developed to analyze them. One such tool is the *step response*, which describes the relationship between an input that changes from zero to a constant value abruptly (a step input) and the corresponding output. The step response is very useful in characterizing the performance of a dynamical system, and it is often used to specify the desired dynamics. A sample step response is shown in Figure 1.12a.

Another way to describe a linear time-invariant system is to represent it by its response to sinusoidal input signals. This is called the *frequency response*, and a rich, powerful theory with many concepts and strong, useful results has emerged for systems that can be described by their frequency response. The results are based on the theory of complex variables and Laplace transforms. The basic idea behind frequency response is that we can completely characterize the behavior of a system by its steady-state response to sinusoidal inputs. Roughly speaking, this is done

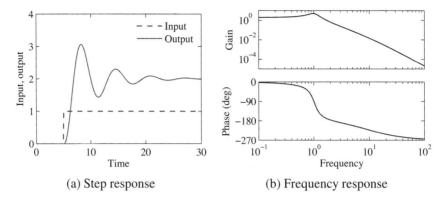

(a) Step response   (b) Frequency response

Figure 1.12: Input/output response of a linear system. The step response (a) shows the output of the system due to an input that changes from 0 to 1 at time $t = 5$ s. The frequency response (b) shows the amplitude gain and phase change due to a sinusoidal input at different frequencies.

by decomposing any arbitrary signal into a linear combination of sinusoids (e.g., by using the Fourier transform) and then using linearity to compute the output by combining the response to the individual frequencies. A sample frequency response is shown in Figure 1.12b.

The input/output view lends itself naturally to experimental determination of system dynamics, where a system is characterized by recording its response to particular inputs, e.g., a step or a set of sinusoids over a range of frequencies.

The control view

When control theory emerged as a discipline in the 1940s, the approach to dynamics was strongly influenced by the electrical engineering (input/output) view. A second wave of developments in control, starting in the late 1950s, was inspired by mechanics, where the state space perspective was used. The emergence of space flight is a typical example, where precise control of the orbit of a spacecraft is essential. These two points of view gradually merged into what is today the state space representation of input/output systems.

The development of state space models involved modifying the models from mechanics to include external actuators and sensors and utilizing more general forms of equations. In control, models often take the form

$$\frac{dx}{dt} = f(x,u), \qquad y = h(x,u), \tag{1.1}$$

where x is a vector of state variables, u is a vector of control signals and y is a vector of measurements. The term dx/dt (sometimes also written as $\dot{x}$) represents the derivative of x with respect to time, now considered a vector, and f and h

are (possibly nonlinear) mappings of their arguments to vectors of the appropriate dimension.

Adding inputs and outputs has increased the richness of the classical problems and led to many new concepts. For example, it is natural to ask if possible states x can be reached with the proper choice of u (reachability) and if the measurement y contains enough information to reconstruct the state (observability). These topics are addressed in greater detail in Åström and Murray [1].

A final development in building the control point of view was the emergence of disturbances and model uncertainty as critical elements in the theory. The simple way of modeling disturbances as deterministic signals like steps and sinusoids has the drawback that such signals cannot be predicted precisely. A more realistic approach is to model disturbances as random signals. This viewpoint gives a natural connection between prediction and control. The dual views of input/output representations and state space representations are particularly useful when modeling uncertainty since state models are convenient to describe a nominal model but uncertainties are easier to describe using input/output models (often via a frequency response description).

An interesting observation in the design of control systems is that feedback systems can often be analyzed and designed based on comparatively simple models. The reason for this is the inherent robustness of feedback systems. However, other uses of models may require more complexity and more accuracy. One example is feedforward control strategies, where one uses a model to precompute the inputs that cause the system to respond in a certain way. Another area is system validation, where one wishes to verify that the detailed response of the system performs as it was designed. Because of these different uses of models, it is common to use a hierarchy of models having different complexity and fidelity.

State space systems

The state of a system is a collection of variables that summarize the past of a system for the purpose of predicting the future. For a biochemical system the state is composed of the variables required to account for the current context of the cell, including the concentrations of the various species and complexes that are present. It may also include the spatial locations of the various molecules. A key issue in modeling is to decide how accurately this information has to be represented. The state variables are gathered in a vector $x \in \mathbb{R}^n$ called the *state vector*. The control variables are represented by another vector $u \in \mathbb{R}^p$, and the measured signal by the vector $y \in \mathbb{R}^q$. A system can then be represented by the differential equation (1.1), where $f : \mathbb{R}^n \times \mathbb{R}^q \to \mathbb{R}^n$ and $h : \mathbb{R}^n \times \mathbb{R}^q \to \mathbb{R}^m$ are smooth mappings. We call a model of this form a *state space model*.

The dimension of the state vector is called the *order* of the system. The system (1.1) is called *time-invariant* because the functions f and h do not depend explicitly on time t; there are more general time-varying systems where the functions do depend on time. The model consists of two functions: the function f gives the rate of change of the state vector as a function of state x and control u, and the function h gives the measured values as functions of state x and control u.

A system is called a *linear* state space system if the functions f and h are linear in x and u. A linear state space system can thus be represented by

$$\frac{dx}{dt} = Ax + Bu, \qquad y = Cx + Du, \qquad (1.2)$$

where A, B, C and D are constant matrices. Such a system is said to be *linear and time-invariant*, or LTI for short. The matrix A is called the *dynamics matrix*, the matrix B is called the *control matrix*, the matrix C is called the *sensor matrix* and the matrix D is called the *direct term*. Frequently systems will not have a direct term, indicating that the control signal does not influence the output directly.

1.5   From systems to synthetic biology

The rapidly growing field of synthetic biology seeks to use biological principles and processes to build useful engineering devices and systems. Applications of synthetic biology range from materials production (drugs, biofuels) to biological sensing and diagnostics (chemical detection, medical diagnostics) to biological machines (bioremediation, nanoscale robotics). Like many other fields at the time of their infancy (electronics, software, networks), it is not yet clear where synthetic biology will have its greatest impact. However, recent advances such as the ability to "boot up" a chemically synthesized genome [31] demonstrate the ability to synthesize systems that offer the possibility of creating devices with substantial functionality. At the same time, the tools and processes available to design systems of this complexity are much more primitive, and *de novo* synthetic circuits typically use a tiny fraction of the number of genetic elements of even the smallest microorganisms [80].

Several scientific and technological developments over the past four decades have set the stage for the design and fabrication of early synthetic biomolecular circuits (see Figure 1.13). An early milestone in the history of synthetic biology can be traced back to the discovery of mathematical logic in gene regulation. In their 1961 paper, Jacob and Monod introduced for the first time the idea of gene expression regulation through transcriptional feedback [48]. Only a few years later (1969), *restriction enzymes* that cut double-stranded DNA at specific recognition sites were discovered by Arber and co-workers [5]. These enzymes were a major enabler of recombinant DNA technology, in which genes from one organism are extracted and spliced into the chromosome of another. One of the most celebrated products of this technology was the large scale production of insulin by employing *E. coli* bacteria as a cell factory [97].

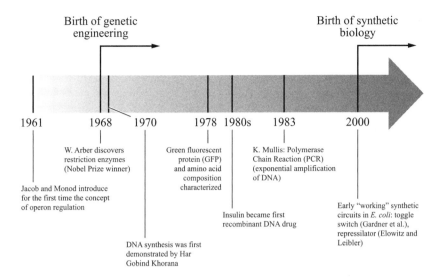

Figure 1.13: Milestones in the history of synthetic biology.

Another key innovation was the development of the polymerase chain reaction (PCR), devised in the 1980s, which allows exponential amplification of small amounts of DNA and can be used to obtain sufficient quantities for use in a variety of molecular biology laboratory protocols where higher concentrations of DNA are required. Using PCR, it is possible to "copy" genes and other DNA sequences out of their host organisms.

The developments of recombinant DNA technology, PCR and artificial synthesis of DNA provided the ability to "cut and paste" natural or synthetic promoters and genes in almost any fashion. This cut and paste procedure is called *cloning* and traditionally consists of four primary steps: fragmentation, *ligation*, *transfection* (or *transformation*) and *screening*. The DNA of interest is first isolated using restriction enzymes and/or PCR amplification. Then, a ligation procedure is employed in which the amplified fragment is inserted into a vector. The vector is often a piece of circular DNA, called a plasmid, that has been linearized by means of restriction enzymes that cleave it at appropriate restriction sites. The vector is then incubated with the fragment of interest with an enzyme called *DNA ligase*, producing a single piece of DNA with the target DNA inserted. The next step is to transfect (or transform) the DNA into living cells, where the natural replication mechanisms of the cell will duplicate the DNA when the cell divides. This process does not transfect all cells, and so a selection procedure is required to isolate those cells that have the desired DNA inserted in them. This is typically done by using a plasmid that gives the cell resistance to a specific antibiotic; cells grown in the presence of that antibiotic will only live if they contain the plasmid. Further selection can be done to ensure that the inserted DNA is also present.

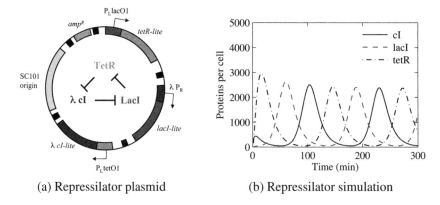

(a) Repressilator plasmid   (b) Repressilator simulation

Figure 1.14: The repressilator genetic regulatory network. (a) A schematic diagram of the repressilator, showing the layout of the genes in the plasmid that holds the circuit as well as the circuit diagram (center). The flat headed arrow between the protein names represents repression. (b) A simulation of a simple model for the repressilator, showing the oscillation of the individual protein concentrations. (Figure courtesy M. Elowitz.)

Once a circuit has been constructed, its performance must be verified and, if necessary, debugged. This is often done with the help of *fluorescent reporters*. The most famous of these is GFP, which was isolated from the jellyfish *Aequorea victoria* in 1978 by Shimomura [88]. Further work by Chalfie and others in the 1990s enabled the use of GFP in *E. coli* as a fluorescent reporter by inserting it into an appropriate point in an artificial circuit [19]. By using spectrofluorometry, fluorescent microscopy or flow cytometry, it is possible to measure the amount of fluorescence in individual cells or collections of cells and characterize the performance of a circuit in the presence of inducers or other factors. Two early examples of the application of these technologies were the *repressilator* [26] and a synthetic genetic switch [30].

The repressilator is a synthetic circuit in which three proteins each repress another in a cycle. This is shown schematically in Figure 1.14a, where the three proteins are TetR, λ cI and LacI. The basic idea of the repressilator is that if TetR is present, then it represses the production of λ cI. If λ cI is absent, then LacI is produced (at the unregulated transcription rate), which in turn represses TetR. Once TetR is repressed, then λ cI is no longer repressed, and so on. If the dynamics of the circuit are designed properly, the resulting protein concentrations will oscillate, as shown in Figure 1.14b.

The repressilator can be constructed using the techniques described above. We can make copies of the individual promoters and genes that form our circuit by using PCR to amplify the selected sequences out of the original organisms in which they were found. TetR is the tetracycline resistance repressor protein that is found in gram-negative bacteria (such as *E. coli*) and is part of the circuitry that provides resistance to tetracycline. LacI is the gene that produces *lac* repressor, responsible

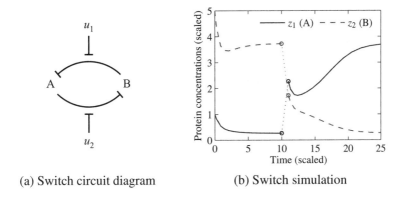

(a) Switch circuit diagram   (b) Switch simulation

Figure 1.15: Stability of a genetic switch. The circuit diagram in (a) represents two proteins that are each repressing the production of the other. The inputs u_1 and u_2 interfere with this repression, allowing the circuit dynamics to be modified. The simulation in (b) shows the time response of the system starting from two different initial conditions. The initial portion of the curve corresponds to protein B having higher concentration than A, and converges to an equilibrium where A is off and B is on. At time $t = 10$, the concentrations are perturbed, moving the concentrations into a region of the state space where solutions converge to the equilibrium point with the A on and B off.

for turning off the *lac* operon in the lactose metabolic pathway in *E. coli*. And λcI comes from λ phage, where it is part of the regulatory circuitry that regulates lysis and lysogeny.

By using restriction enzymes and related techniques, we can separate the natural promoters from their associated genes, and then ligate (reassemble) them in a new order and insert them into a "backbone" vector (the rest of the plasmid, including the origin of replication and appropriate antibiotic resistance). This DNA is then transformed into cells that are grown in the presence of an antibiotic, so that only those cells that contain the repressilator can replicate. Finally, we can take individual cells containing our circuit and let them grow under a microscope to image fluorescent reporters coupled to the oscillator.

Another early circuit in the synthetic biology toolkit is a genetic switch built by Gardner et al. [30]. The genetic switch consists of two repressors connected together in a cycle, as shown in Figure 1.15a. The intuition behind this circuit is that if the gene A is being expressed, it will repress production of B and maintain its expression level (since the protein corresponding to B will not be present to repress A). Similarly, if B is being expressed, it will repress the production of A and maintain its expression level. This circuit thus implements a type of *bistability* that can be used as a simple form of memory. Figure 1.15b shows the time traces for the system, illustrating the bistable nature of the circuit. When the initial condition starts with a concentration of protein B greater than that of A, the solution converges to the equilibrium point where B is on and A is off. If A is greater than

B, then the opposite situation results.

These seemingly simple circuits took years of effort to get to work, but showed that it was possible to synthesize a biological circuit that performed a desired function that was not originally present in a natural system. Today, commercial synthesis of DNA sequences and genes has become cheaper and faster, with a price often below $0.20 per base pair.[3] The combination of inexpensive synthesis technologies, new advances in cloning techniques, and improved devices for imaging and measurement has vastly simplified the process of producing a sequence of DNA that encodes a given set of genes, operator sites, promoters and other functions. These techniques are a routine part of undergraduate courses in molecular and synthetic biology.

As illustrated by the examples above, current techniques in synthetic biology have demonstrated the ability to program biological function by designing DNA sequences that implement simple circuits. Most current devices make use of transcriptional or post-transcriptional processing, resulting in very slow time scales (response times typically measured in tens of minutes to hours). This restricts their use in systems where faster response to environmental signals is needed, such as rapid detection of a chemical signal or fast response to changes in the internal environment of the cell. In addition, existing methods for biological circuit design have limited modularity (reuse of circuit elements requires substantial redesign or tuning) and typically operate in very narrow operating regimes (e.g., a single species grown in a single type of media under carefully controlled conditions). Furthermore, engineered circuits inserted into cells can interact with the host organism and have other unintended interactions.

As an illustration of the dynamics of synthetic devices, Figure 1.16 shows a typical response of a genetic element to an inducer molecule [18]. In this circuit, an external signal of homoserine lactone (HSL) is applied at time zero and the system reaches 10% of the steady state value in approximately 15 minutes. This response is limited in part by the time required to synthesize the output protein (GFP), including delays due to transcription, translation and folding. Since this is the response time for the underlying "actuator," circuits that are composed of feedback interconnections of such genetic elements will typically operate at 5–10 times slower speeds. While these speeds are appropriate in many applications (e.g., regulation of steady state enzyme levels for materials production), in the context of biochemical sensors or systems that must maintain a steady operating point in more rapidly changing thermal or chemical environments, this response time is too slow to be used as an effective engineering approach.

By comparison, the input/output response for the signaling component in *E. coli* chemotaxis is shown in Figure 1.17 [87]. Here the response of the kinase CheA is plotted in response to an exponential ramp in the ligand concentration. The response is extremely rapid, with the time scale measured in seconds. This rapid

---

[3]As of this writing; divide by a factor of two for every two years after the publication date.

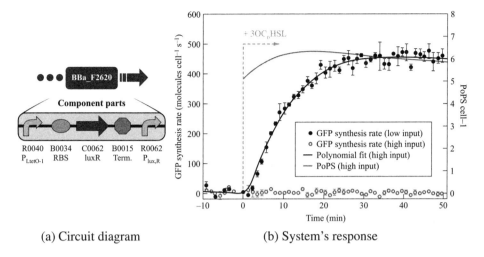

(a) Circuit diagram   (b) System's response

Figure 1.16: Expression of a protein using an inducible promoter [18]. (a) The circuit diagram indicates the DNA sequences that are used to construct the part (chosen from the BioBrick library). (b) The measured response of the system to a step change in the inducer level (HSL). Figures adapted from [18].

response is implemented by conformational changes in the proteins involved in the circuit, rather than regulation of transcription or other slower processes.

The field of synthetic biology has the opportunity to provide new approaches to solving engineering and scientific problems. Sample engineering applications include the development of synthetic circuits for producing biofuels, ultrasensitive chemical sensors, or production of materials with specific properties that are tuned to commercial needs. In addition to the potential impact on new biologically engineered devices, there is also the potential for impact in improved understanding of biological processes. For example, many diseases such as cancer and Parkinson's

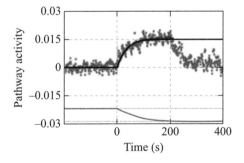

Figure 1.17: Responses of *E. coli* chemotaxis signaling network to exponential ramps in ligand concentration. Time responses of the "sensing" subsystem (from Shimizu, Tu and Berg [87]), showing the response to exponential inputs. Figure adapted from [87].

disease are closely tied to kinase dysfunction. The analysis of robust systems of kinases and the ability to synthesize systems that support or invalidate biological hypotheses may lead to a better systems understanding of failure modes that lead to such diseases.

1.6 Further reading

There are numerous survey articles and textbooks that provide more detailed introductions to the topics introduced in this chapter. In the field of systems biology, the textbook by Alon [4] provides a broad view of some of the key elements of modern systems biology. A more comprehensive set of topics is covered in the textbook by Klipp [56], while a more engineering-oriented treatment of modeling of biological circuits can be found in the text by Myers [74]. Two other books that are particularly noteworthy are Ptashne's book on the phage λ [79] and Madhani's book on yeast [63], both of which use well-studied model systems to describe a general set of mechanisms and principles that are present in many different types of organisms.

Several textbooks and research monographs provide excellent resources for modeling and analysis of biomolecular dynamics and regulation. J. D. Murray's two-volume text [73] on biological modeling is an excellent reference with many examples of biomolecular dynamics. The textbook by Phillips, Kondev and Theriot [78] provides a quantitative approach to understanding biological systems, including many of the concepts discussed in this chapter. Courey [20] gives a detailed description of mechanisms transcriptional regulation. The topics in dynamical systems and control theory that are briefly introduced here are covered in more detail in Åström and Murray [1] and can also be found in the text by Ellner and Guckenheimer [25].

Synthetic biology is a rapidly evolving field that includes many different subareas of research, but few textbooks are currently available. In the specific area of biological circuit design that we focus on here, there are a number of good survey and review articles. The article by Baker et al. [8] provides a high level description of the basic approach and opportunities. Additional survey and review papers include Voigt [99], Purnick and Weiss [80], and Khalil and Collins [54].

Chapter 2
Dynamic Modeling of Core Processes

The goal of this chapter is to describe basic biological mechanisms in a way that can be represented by simple dynamical models. We begin the chapter with a discussion of the basic modeling formalisms that we will utilize to model biomolecular feedback systems. We then proceed to study a number of core processes within the cell, providing different model-based descriptions of the dynamics that will be used in later chapters to analyze and design biomolecular systems. The focus in this chapter and the next is on deterministic models using ordinary differential equations; Chapter 4 describes how to model the stochastic nature of biomolecular systems.

2.1 Modeling chemical reactions

In order to develop models for some of the core processes of the cell, we will need to build up a basic description of the biochemical reactions that take place, including production and degradation of proteins, regulation of transcription and translation, and intracellular sensing, action and computation. As in other disciplines, biomolecular systems can be modeled in a variety of different ways, at many different levels of resolution, as illustrated in Figure 2.1. The choice of which model to use depends on the questions that we want to answer, and good modeling takes practice, experience, and iteration. We must properly capture the aspects of the system that are important, reason about the appropriate temporal and spatial scales to be included, and take into account the types of simulation and analysis tools to be applied. Models used for analyzing existing systems should make testable predictions and provide insight into the underlying dynamics. Design models must additionally capture enough of the important behavior to allow decisions regarding how to interconnect subsystems, choose parameters and design regulatory elements.

In this section we describe some of the basic modeling frameworks that we will build on throughout the rest of the text. We begin with brief descriptions of the relevant physics and chemistry of the system, and then quickly move to models that focus on capturing the behavior using reaction rate equations. In this chapter our emphasis will be on dynamics with time scales measured in seconds to hours and mean behavior averaged across a large number of molecules. We touch only briefly on modeling in the case where stochastic behavior dominates and defer a more detailed treatment until Chapter 4.

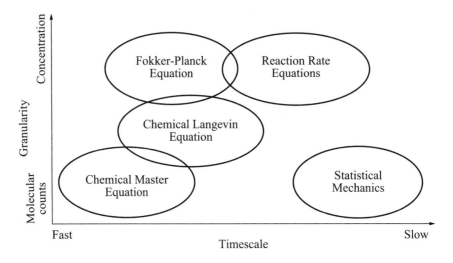

Figure 2.1: Different methods of modeling biomolecular systems.

Reaction kinetics

At the fine end of the modeling scale, we can attempt to model the *molecular dynamics* of the cell, in which we attempt to model the individual proteins and other species and their interactions via molecular-scale forces and motions. At this scale, the individual interactions between protein domains, DNA and RNA are resolved, resulting in a highly detailed model of the dynamics of the cell.

For our purposes in this text, we will not require the use of such a detailed scale and we will consider the main modeling formalisms depicted in Figure 2.1. We start with the abstraction of molecules that interact with each other through stochastic events that are guided by the laws of thermodynamics. We begin with an equilibrium point of view, commonly referred to as *statistical mechanics*, and then briefly describe how to model the (statistical) dynamics of the system using chemical kinetics. We cover both of these points of view very briefly here, primarily as a stepping stone to deterministic models.

The underlying representation for both statistical mechanics and chemical kinetics is to identify the appropriate *microstates* of the system. A microstate corresponds to a given configuration of the components (species) in the system relative to each other and we must enumerate all possible configurations between the molecules that are being modeled.

As an example, consider the distribution of RNA polymerase in the cell. It is known that most RNA polymerases are bound to the DNA in a cell, either as they produce RNA or as they diffuse along the DNA in search of a promoter site. Hence we can model the microstates of the RNA polymerase system as all possible locations of the RNA polymerase in the cell, with the vast majority of these corresponding to the RNA polymerase at some location on the DNA. This is illustrated

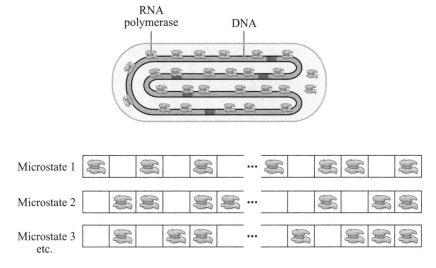

Figure 2.2: Microstates for RNA polymerase. Each microstate of the system corresponds to the RNA polymerase being located at some position in the cell. If we discretize the possible locations on the DNA and in the cell, the microstate corresponds to all possible non-overlapping locations of the RNA polymerases. Figure adapted from Phillips, Kondev and Theriot [78].

in Figure 2.2. In statistical mechanics, we model the configuration of the cell by the probability that the system is in a given microstate. This probability can be calculated based on the energy levels of the different microstates. The laws of statistical mechanics state that if we have a set of microstates Q, then the steady state probability that the system is in a particular microstate q is given by

$$\mathbb{P}(q) = \frac{1}{Z}e^{-E_q/(k_BT)}, \tag{2.1}$$

where E_q is the energy associated with the microstate $q \in Q$, k_B is the Boltzmann constant, T is the temperature in degrees Kelvin, and Z is a normalizing factor, known as the *partition function*,

$$Z = \sum_{q \in Q} e^{-E_q/(k_BT)}.$$

By keeping track of those microstates that correspond to a given system state (also called a *macrostate*), we can compute the overall probability that a given macrostate is reached. Thus, if we have a set of states $S \subset Q$ that corresponds to a given macrostate, then the probability of being in the set S is given by

$$P(S) = \frac{1}{Z} \sum_{q \in S} e^{-E_q/(k_BT)} = \frac{\sum_{q \in S} e^{-E_q/(k_BT)}}{\sum_{q \in Q} e^{-E_q/(k_BT)}}. \tag{2.2}$$

This can be used, for example, to compute the probability that some RNA polymerase is bound to a given promoter, averaged over many independent samples, and from this we can reason about the rate of expression of the corresponding gene.

Statistical mechanics describes the steady state distribution of microstates, but does not tell us how the microstates evolve in time. To include the dynamics, we must consider the *chemical kinetics* of the system and model the probability that we transition from one microstate to another in a given period of time. Let q represent the microstate of the system, which we shall take as a vector of integers that represents the number of molecules of a specific type (species) in given configurations or locations. Assume we have a set of m chemical reactions $\texttt{Rj}$, $j = 1, \ldots, M$, in which a chemical reaction is a process that leads to the transformation of one set of chemical species to another one. We use ξ_j to represent the change in state q associated with reaction $\texttt{Rj}$. We describe the kinetics of the system by making use of the *propensity function* $a_j(q,t)$ associated with reaction $\texttt{Rj}$, which captures the instantaneous probability that at time t a system will transition between state q and state $q + \xi_j$.

More specifically, the propensity function is defined such that

$$
\begin{aligned}
a_j(q,t)dt \quad = \quad &\text{Probability that reaction } \texttt{Rj} \text{ will occur between time } t \\
&\text{and time } t + dt \text{ given that the microstate is } q.
\end{aligned}
$$

We will give more detail in Chapter 4 regarding the validity of this functional form, but for now we simply assume that such a function can be defined for our system.

Using the propensity function, we can keep track of the probability distribution for the state by looking at all possible transitions into and out of the current state. Specifically, given $P(q,t)$, the probability of being in state q at time t, we can compute the time derivative $dP(q,t)/dt$ as

$$
\frac{dP}{dt}(q,t) = \sum_{j=1}^{M} \Big(a_j(q - \xi_j, t) P(q - \xi_j, t) - a_j(q,t) P(q,t) \Big). \tag{2.3}
$$

This equation (and its variants) is called the *chemical master equation* (CME). The first sum on the right-hand side represents the transitions into the state q from some other state $q - \xi_j$ and the second sum represents the transitions out of the state q.

The dynamics of the distribution $P(q,t)$ depend on the form of the propensity functions $a_j(q,t)$. Consider a simple reversible reaction of the form

$$
\text{A} + \text{B} \rightleftharpoons \text{AB}, \tag{2.4}
$$

in which a molecule of A and a molecule of B come together to form the complex AB, in which A and B are bound to each other, and this complex can, in turn, dissociate back into the A and B species. In the sequel, to make notation easier, we will sometimes represent the complex AB as A : B. It is often useful to write reversible reactions by splitting the forward reaction from the backward reaction:

$$
\begin{aligned}
\texttt{Rf:} \quad & A + B \longrightarrow AB, \\
\texttt{Rr:} \quad & AB \longrightarrow A + B.
\end{aligned}
\tag{2.5}
$$

We assume that the reaction takes place in a well-stirred volume Ω and let the configurations q be represented by the number of each species that is present. The forward reaction $\texttt{Rf}$ is a bimolecular reaction and we will see in Chapter 4 that it has a propensity function

$$
a_f(q) = \frac{k_f}{\Omega} n_A n_B,
$$

where k_f is a parameter that depends on the forward reaction, and n_A and n_B are the number of molecules of each species. The reverse reaction $\texttt{Rr}$ is a unimolecular reaction and we will see that it has a propensity function

$$
a_r(q) = k_r n_{AB},
$$

where k_r is a parameter that depends on the reverse reaction and n_{AB} is the number of molecules of AB that are present.

If we now let $q = (n_A, n_B, n_{AB})$ represent the microstate of the system, then we can write the chemical master equation as

$$
\frac{dP}{dt}(n_A, n_B, n_{AB}) = k_r n_{AB} P(n_A - 1, n_B - 1, n_{AB} + 1) - \frac{k_f}{\Omega} n_A n_B P(n_A, n_B, n_{AB}).
$$

The first term on the right-hand side represents the transitions into the microstate $q = (n_A, n_B, n_{AB})$ and the second term represents the transitions out of that state.

The number of differential equations depends on the number of molecules of A, B and AB that are present. For example, if we start with one molecule of A, one molecule of B, and three molecules of AB, then the possible states and dynamics are

$$
\begin{aligned}
q_0 &= (1,0,4), & dP_0/dt &= 3k_r P_1, \\
q_1 &= (2,1,3), & dP_1/dt &= 4k_r P_0 - 2(k_f/\Omega) P_1, \\
q_2 &= (3,2,2), & dP_2/dt &= 3k_r P_1 - 6(k_f/\Omega) P_2, \\
q_3 &= (4,3,1), & dP_3/dt &= 2k_r P_2 - 12(k_f/\Omega) P_3, \\
q_4 &= (5,4,0), & dP_4/dt &= 1k_r P_3 - 20(k_f/\Omega) P_4,
\end{aligned}
$$

where $P_i = P(q_i, t)$. Note that the states of the chemical master equation are the probabilities that we are in a specific microstate, and the chemical master equation is a *linear* differential equation (we see from equation (2.3) that this is true in general).

The primary difference between the statistical mechanics description given by equation (2.1) and the chemical kinetics description in equation (2.3) is that the master equation formulation describes how the probability of being in a given microstate evolves over time. Of course, if the propensity functions and energy levels are modeled properly, the steady state, average probabilities of being in a given microstate, should be the same for both formulations.

Reaction rate equations

Although very general in form, the chemical master equation suffers from being a very high-dimensional representation of the dynamics of the system. We shall see in Chapter 4 how to implement simulations that obey the master equation, but in many instances we will not need this level of detail in our modeling. In particular, there are many situations in which the number of molecules of a given species is such that we can reason about the behavior of a chemically reacting system by keeping track of the *concentration* of each species as a real number. This is of course an approximation, but if the number of molecules is sufficiently large, then the approximation will generally be valid and our models can be dramatically simplified.

To go from the chemical master equation to a simplified form of the dynamics, we begin by making a number of assumptions. First, we assume that we can represent the state of a given species by its concentration n_A/Ω, where n_A is the number of molecules of A in a given volume Ω. We also treat this concentration as a real number, ignoring the fact that the real concentration is quantized. Finally, we assume that our reactions take place in a well-stirred volume, so that the rate of interactions between two species is solely determined by the concentrations of the species.

Before proceeding, we should recall that in many (and perhaps most) situations inside of cells, these assumptions are *not* particularly good ones. Biomolecular systems often have very small molecular counts and are anything but well mixed. Hence, we should not expect that models based on these assumptions should perform well at all. However, experience indicates that in many cases the basic form of the equations provides a good model for the underlying dynamics and hence we often find it convenient to proceed in this manner.

Putting aside our potential concerns, we can now create a model for the dynamics of a system consisting of a set of species S_i, $i = 1,\ldots,n$, undergoing a set of reactions R_j, $j = 1,\ldots,M$. We write $x_i = [S_i] = n_{S_i}/\Omega$ for the concentration of species i (viewed as a real number). Because we are interested in the case where the number of molecules is large, we no longer attempt to keep track of every possible configuration, but rather simply assume that the state of the system at any given time is given by the concentrations x_i. Hence the state space for our system is given by $x \in \mathbb{R}^n$ and we seek to write our dynamics in the form of an ordinary differential equation (ODE)

$$\frac{dx}{dt} = f(x,\theta),$$

where $\theta \in \mathbb{R}^p$ represents the vector of parameters that govern dynamic behavior and $f : \mathbb{R}^n \times \mathbb{R}^p \to \mathbb{R}^n$ describes the rate of change of the concentrations as a function of the instantaneous concentrations and parameter values.

To illustrate the general form of the dynamics, we consider again the case of a basic bimolecular reaction

$$A + B \rightleftharpoons AB.$$

Each time the forward reaction occurs, we decrease the number of molecules of A and B by one and increase the number of molecules of AB (a separate species) by one. Similarly, each time the reverse reaction occurs, we decrease the number of molecules of AB by one and increase the number of molecules of A and B.

Using our discussion of the chemical master equation, we know that the likelihood that the forward reaction occurs in a given interval dt is given by $a_f(q)dt = (k_f/\Omega)n_A n_B dt$ and the reverse reaction has likelihood $a_r(q) = k_r n_{AB}$. If we assume that n_{AB} is a real number instead of an integer and ignore some of the formalities of random variables, we can describe the evolution of n_{AB} using the equation

$$n_{AB}(t + dt) = n_{AB}(t) + a_f(q - \xi_f)dt - a_r(q)dt.$$

Here we let q be the state of the system with the number of molecules of AB equal to n_{AB} and ξ_f represents the change in state from the forward reaction (n_A and n_B are decreased by one and n_{AB} is increased by one). Roughly speaking, this equation states that the (approximate) number of molecules of AB at time $t + dt$ compared with time t increases by the probability that the forward reaction occurs in time dt and decreases by the probability that the reverse reaction occurs in that period.

To convert this expression into an equivalent one for the concentration of the species AB, we write $[AB] = n_{AB}/\Omega$, $[A] = n_A/\Omega$, $[B] = n_B/\Omega$, and substitute the expressions for $a_f(q)$ and $a_r(q)$:

$$\begin{aligned}
[AB](t + dt) - [AB](t) &= \left(a_f(q - \xi_f, t) - a_r(q) \right)/\Omega \cdot dt \\
&= \left(k_f n_A n_B /\Omega^2 - k_r n_{AB}/\Omega \right) dt \\
&= \left(k_f [A][B] - k_r [AB] \right) dt.
\end{aligned}$$

Taking the limit as dt approaches zero, we obtain

$$\frac{d}{dt}[AB] = k_f[A][B] - k_r[AB].$$

Our derivation here has skipped many important steps, including a careful derivation using random variables and some assumptions regarding the way in which dt approaches zero. These are described in more detail when we derive the chemical Langevin equation (CLE) in Chapter 4, but the basic form of the equations are correct under the assumptions that the reactions are well-stirred and the molecular counts are sufficiently large.

In a similar fashion we can write equations to describe the dynamics of A and B and the entire system of equations is given by

$$\frac{d[A]}{dt} = k_r[AB] - k_f[A][B], \qquad\qquad \frac{dA}{dt} = k_rC - k_fA \cdot B,$$

$$\frac{d[B]}{dt} = k_r[AB] - k_f[A][B], \quad \text{or} \quad \frac{dB}{dt} = k_rC - k_fA \cdot B,$$

$$\frac{d[AB]}{dt} = k_f[A][B] - k_r[AB], \qquad\qquad \frac{dC}{dt} = k_fA \cdot B - k_rC,$$

where $C = [AB]$, $A = [A]$, and $B = [B]$. These equations are known as the *mass action kinetics* or the *reaction rate equations* for the system. The parameters k_f and k_r are called the *rate constants* and they match the parameters that were used in the underlying propensity functions.

Note that the same rate constants appear in each term, since the rate of production of AB must match the rate of depletion of A and B and vice versa. We adopt the standard notation for chemical reactions with specified rates and write the individual reactions as

$$A + B \xrightarrow{k_f} AB, \qquad AB \xrightarrow{k_r} A + B,$$

where k_f and k_r are the reaction rate constants. For bidirectional reactions we can also write

$$A + B \underset{k_r}{\overset{k_f}{\rightleftharpoons}} AB.$$

It is easy to generalize these dynamics to more complex reactions. For example, if we have a reversible reaction of the form

$$A + 2B \underset{k_r}{\overset{k_f}{\rightleftharpoons}} 2C + D,$$

where A, B, C and D are appropriate species and complexes, then the dynamics for the species concentrations can be written as

$$\frac{d}{dt}A = k_rC^2 \cdot D - k_fA \cdot B^2, \qquad \frac{d}{dt}C = 2k_fA \cdot B^2 - 2k_rC^2 \cdot D,$$

$$\frac{d}{dt}B = 2k_rC^2 \cdot D - 2k_fA \cdot B^2, \qquad \frac{d}{dt}D = k_fA \cdot B^2 - k_rC^2 \cdot D. \tag{2.6}$$

Rearranging this equation, we can write the dynamics as

$$\frac{d}{dt}\begin{pmatrix} A \\ B \\ C \\ D \end{pmatrix} = \begin{pmatrix} -1 & 1 \\ -2 & 2 \\ 2 & -2 \\ 1 & -1 \end{pmatrix} \begin{pmatrix} k_fA \cdot B^2 \\ k_rC^2 \cdot D \end{pmatrix}. \tag{2.7}$$

We see that in this decomposition, the first term on the right-hand side is a matrix of integers reflecting the stoichiometry of the reactions and the second term is a vector of rates of the individual reactions.

More generally, given a chemical reaction consisting of a set of species S_i, $i = 1, \ldots, n$ and a set of reactions Rj, $j = 1, \ldots, M$, we can write the mass action kinetics in the form

$$\frac{dx}{dt} = Nv(x),$$

where $N \in \mathbb{R}^{n \times M}$ is the *stoichiometry matrix* for the system and $v(x) \in \mathbb{R}^M$ is the *reaction flux vector*. Each row of $v(x)$ corresponds to the rate at which a given reaction occurs and the corresponding column of the stoichiometry matrix corresponds to the changes in concentration of the relevant species. For example, for the system in equation (2.7) we have

$$x = (A, B, C, D), \qquad N = \begin{pmatrix} -1 & 1 \\ -2 & 2 \\ 2 & -2 \\ 1 & -1 \end{pmatrix}, \qquad v(x) = \begin{pmatrix} k_f A \cdot B^2 \\ k_r C^2 \cdot D \end{pmatrix}.$$

The conservation of species is at the basis of reaction rate models since species are usually transformed, but are not created from nothing or destroyed. Even the basic process of protein degradation transforms a protein of interest A into a product X that is not used in any other reaction. Specifically, the degradation rate of a protein is determined by the amounts of proteases present, which bind to recognition sites (degradation tags) and then degrade the protein. Degradation of a protein A by a protease P can then be modeled by the following two-step reaction:

$$A + P \underset{d}{\overset{a}{\rightleftharpoons}} AP \overset{k}{\rightarrow} P + X.$$

As a result of the reaction, protein A has "disappeared," so that this reaction is often simplified to $A \longrightarrow \emptyset$. Similarly, the birth of a molecule is a complicated process that involves many reactions and species, as we will see later in this chapter. When the process that creates a species of interest A is not relevant for the problem under study, we will use the shorter description of a birth reaction given by

$$\emptyset \overset{k_f}{\longrightarrow} A$$

and describe its dynamics using the differential equation

$$\frac{dA}{dt} = k_f.$$

Example 2.1 (Covalent modification of a protein). Consider the set of reactions involved in the phosphorylation of a protein by a kinase, as shown in Figure 2.3.

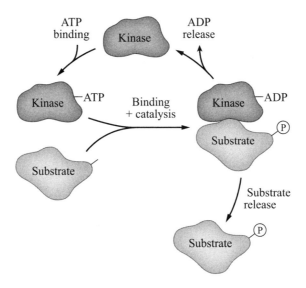

Figure 2.3: Phosphorylation of a protein via a kinase. In the process of phosphorylation, a protein called a kinase binds to ATP (adenosine triphosphate) and transfers one of the phosphate groups (P) from ATP to a substrate, hence producing a phosphorylated substrate and ADP (adenosine diphosphate). Figure adapted from Madhani [63].

Let S represent the substrate, K represent the kinase and S^* represent the phosphorylated (activated) substrate. The sets of reactions illustrated in Figure 2.3 are

$$R1: \quad K + ATP \longrightarrow K{:}ATP, \qquad\qquad R5: \quad S{:}K{:}ATP \longrightarrow S^*{:}K{:}ADP,$$
$$R2: \quad K{:}ATP \longrightarrow K + ATP, \qquad\qquad R6: \quad S^*{:}K{:}ADP \longrightarrow S^* + K{:}ADP,$$
$$R3: \quad S + K{:}ATP \longrightarrow S{:}K{:}ATP, \qquad R7: \quad K{:}ADP \longrightarrow K + ADP,$$
$$R4: \quad S{:}K{:}ATP \longrightarrow S + K{:}ATP, \qquad R8: \quad K + ADP \longrightarrow K{:}ADP.$$

We now write the kinetics for each reaction:

$$v_1 = k_1\,[K][ATP], \qquad\qquad v_5 = k_5\,[S{:}K{:}ATP],$$
$$v_2 = k_2\,[K{:}ATP], \qquad\qquad v_6 = k_6\,[S^*{:}K{:}ADP],$$
$$v_3 = k_3\,[S][K{:}ATP], \qquad\qquad v_7 = k_7\,[K{:}ADP],$$
$$v_4 = k_4\,[S{:}K{:}ATP], \qquad\qquad v_8 = k_8\,[K][ADP].$$

We treat [ATP] as a constant (regulated by the cell) and hence do not directly track its concentration. (If desired, we could similarly ignore the concentration of ADP since we have chosen not to include the many additional reactions in which it participates.)

The kinetics for each species are thus given by

$$\frac{d}{dt}[K] = -v_1 + v_2 + v_7 - v_8, \qquad \frac{d}{dt}[S^*] = v_6,$$

$$\frac{d}{dt}[K{:}ATP] = v_1 - v_2 - v_3 + v_4, \qquad \frac{d}{dt}[S^*{:}K{:}ADP] = v_5 - v_6,$$

$$\frac{d}{dt}[S] = -v_3 + v_4, \qquad \frac{d}{dt}[ADP] = v_7 - v_8,$$

$$\frac{d}{dt}[S{:}K{:}ATP] = v_3 - v_4 - v_5, \qquad \frac{d}{dt}[K{:}ADP] = v_6 - v_7 + v_8.$$

Collecting these equations together and writing the state as a vector, we obtain

$$\frac{d}{dt}\underbrace{\begin{pmatrix} [K] \\ [K{:}ATP] \\ [S] \\ [S{:}K{:}ATP] \\ [S^*] \\ [S^*{:}K{:}ADP] \\ [ADP] \\ [K{:}ADP] \end{pmatrix}}_{x} = \underbrace{\begin{pmatrix} -1 & 1 & 0 & 0 & 0 & 0 & 1 & -1 \\ 1 & -1 & -1 & 1 & 0 & 0 & 0 & 0 \\ 0 & 0 & -1 & 1 & 0 & 0 & 0 & 0 \\ 0 & 0 & 1 & -1 & -1 & 0 & 0 & 0 \\ 0 & 0 & 0 & 0 & 0 & 1 & 0 & 0 \\ 0 & 0 & 0 & 0 & 1 & -1 & 0 & 0 \\ 0 & 0 & 0 & 0 & 0 & 0 & 1 & -1 \\ 0 & 0 & 0 & 0 & 0 & 1 & -1 & 1 \end{pmatrix}}_{N} \underbrace{\begin{pmatrix} v_1 \\ v_2 \\ v_3 \\ v_4 \\ v_5 \\ v_6 \\ v_7 \\ v_8 \end{pmatrix}}_{v(x)},$$

which is in standard stoichiometric form.   ▽

Reduced-order mechanisms

In this section, we look at the dynamics of some common reactions that occur in biomolecular systems. Under some assumptions on the relative rates of reactions and concentrations of species, it is possible to derive reduced-order expressions for the dynamics of the system. We focus here on an informal derivation of the relevant results, but return to these examples in the next chapter to illustrate that the same results can be derived using a more formal and rigorous approach.

Simple binding reaction. Consider the reaction in which two species A and B bind reversibly to form a complex C = AB:

$$A + B \underset{d}{\overset{a}{\rightleftharpoons}} C, \tag{2.8}$$

where a is the *association rate constant* and d is the *dissociation rate constant*. Assume that B is a species that is controlled by other reactions in the cell and that the total concentration of A is conserved, so that $A + C = [A] + [AB] = A_{\text{tot}}$. If the dynamics of this reaction are fast compared to other reactions in the cell, then the amount of A and C present can be computed as a (steady state) function of the amount of B.

To compute how A and C depend on the concentration of B at the steady state, we must solve for the equilibrium concentrations of A and C. The rate equation for C is given by

$$\frac{dC}{dt} = aB \cdot A - dC = aB \cdot (A_{\text{tot}} - C) - dC.$$

By setting $dC/dt = 0$ and letting $K_{\text{d}} := d/a$, we obtain the expressions

$$C = \frac{A_{\text{tot}}(B/K_{\text{d}})}{1 + (B/K_{\text{d}})}, \qquad A = \frac{A_{\text{tot}}}{1 + (B/K_{\text{d}})}.$$

The constant K_{d} is called the *dissociation constant* of the reaction. Its inverse measures the affinity of A binding to B. The steady state value of C increases with B while the steady state value of A decreases with B as more of A is found in the complex C.

Note that when $B \approx K_{\text{d}}$, A and C have equal concentration. Thus the higher the value of K_{d}, the more B is required for A to form the complex C. K_{d} has the units of concentration and it can be interpreted as the concentration of B at which half of the total number of molecules of A are associated with B. Therefore a high K_{d} represents a weak affinity between A and B, while a low K_{d} represents a strong affinity.

Cooperative binding reaction. Assume now that B binds to A only after dimerization, that is, only after binding another molecule of B. Then, we have that reactions (2.8) become

$$B + B \underset{k_2}{\overset{k_1}{\rightleftharpoons}} B_2, \qquad B_2 + A \underset{d}{\overset{a}{\rightleftharpoons}} C, \qquad A + C = A_{\text{tot}},$$

in which $B_2 = B : B$ represents the dimer of B, that is, the complex of two molecules of B bound to each other. The corresponding ODE model is given by

$$\frac{dB_2}{dt} = k_1 B^2 - k_2 B_2 - aB_2 \cdot (A_{\text{tot}} - C) + dC, \qquad \frac{dC}{dt} = aB_2 \cdot (A_{\text{tot}} - C) - dC.$$

By setting $dB_2/dt = 0$, $dC/dt = 0$, and by defining $K_m := k_2/k_1$, we obtain that

$$B_2 = B^2/K_m, \qquad C = \frac{A_{\text{tot}}(B_2/K_{\text{d}})}{1 + (B_2/K_{\text{d}})}, \qquad A = \frac{A_{\text{tot}}}{1 + (B_2/K_{\text{d}})},$$

so that

$$C = \frac{A_{\text{tot}}B^2/(K_m K_{\text{d}})}{1 + B^2/(K_m K_{\text{d}})}, \qquad A = \frac{A_{\text{tot}}}{1 + B^2/(K_m K_{\text{d}})}.$$

As an exercise (Exercise 2.2), the reader can verify that if B binds to A as a complex of n copies of B, that is,

$$B + B + \cdots + B \underset{k_2}{\overset{k_1}{\rightleftharpoons}} B_n, \qquad B_n + A \underset{d}{\overset{a}{\rightleftharpoons}} C, \qquad A + C = A_{\text{tot}},$$

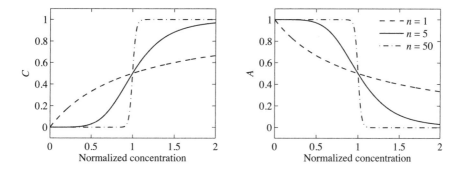

Figure 2.4: Steady state concentrations of the complex C and of A as functions of the concentration of B.

then we have that the expressions of C and A change to

$$C = \frac{A_{\text{tot}}B^n/(K_mK_d)}{1+B^n/(K_mK_d)}, \qquad A = \frac{A_{\text{tot}}}{1+B^n/(K_mK_d)}.$$

In this case, we say that the binding of B to A is *cooperative* with cooperativity n. Figure 2.4 shows the above functions, which are often referred to as Hill functions and n is called the Hill coefficient.

Another type of cooperative binding is when a species R can bind A only after another species B has bound A. In this case, the reactions are given by

$$B + A \underset{d}{\overset{a}{\rightleftharpoons}} C, \qquad R + C \underset{d'}{\overset{a'}{\rightleftharpoons}} C', \qquad A + C + C' = A_{\text{tot}}.$$

Proceeding as above by writing the ODE model and equating the time derivatives to zero to obtain the equilibrium, we obtain the equilibrium relations

$$C = \frac{1}{K_d}B(A_{\text{tot}} - C - C'), \qquad C' = \frac{1}{K_d'K_d}R(A_{\text{tot}} - C - C').$$

By solving this system of two equations for the unknowns C' and C, we obtain

$$C' = \frac{A_{\text{tot}}(B/K_d)(R/K_d')}{1+(B/K_d)+(B/K_d)(R/K_d')}, \qquad C = \frac{A_{\text{tot}}(B/K_d)}{1+(B/K_d)+(B/K_d)(R/K_d')}.$$

In the case in which B would bind cooperatively with other copies of B with cooperativity n, the above expressions become

$$C' = \frac{A_{\text{tot}}(B^n/K_mK_d)(R/K_d')}{1+(B^n/K_mK_d)(R/K_d')+(B^n/K_mK_d)},$$

$$C = \frac{A_{\text{tot}}(B^n/K_mK_d)}{1+(B^n/K_mK_d)(R/K_d')+(B^n/K_mK_d)}.$$

Competitive binding reaction. Finally, consider the case in which two species B_a and B_r both bind to A competitively, that is, they cannot be bound to A at the same time. Let C_a be the complex formed between B_a and A and let C_r be the complex formed between B_r and A. Then, we have the following reactions

$$B_a + A \underset{d}{\overset{a}{\rightleftharpoons}} C_a, \qquad B_r + A \underset{d'}{\overset{a'}{\rightleftharpoons}} C_r, \qquad A + C_a + C_r = A_{tot},$$

for which we can write the differential equation model as

$$\frac{dC_a}{dt} = aB_a \cdot (A_{tot} - C_a - C_r) - dC_a, \qquad \frac{dC_r}{dt} = a'B_r \cdot (A_{tot} - C_a - C_r) - d'C_r.$$

By setting the time derivatives to zero, we obtain

$$C_a(aB_a + d) = aB_a(A_{tot} - C_r), \qquad C_r(a'B_r + d') = a'B_r(A_{tot} - C_a),$$

so that

$$C_r = \frac{B_r(A_{tot} - C_a)}{B_r + K'_d}, \qquad C_a\left(B_a + K_d - \frac{B_aB_r}{B_r + K'_d}\right) = B_a\left(\frac{K'_d}{B_r + K'_d}\right)A_{tot},$$

from which we finally determine that

$$C_a = \frac{A_{tot}(B_a/K_d)}{1 + (B_a/K_d) + (B_r/K'_d)}, \qquad C_r = \frac{A_{tot}(B_r/K'_d)}{1 + (B_a/K_d) + (B_r/K'_d)}.$$

In this derivation, we have assumed that both B_a and B_r bind A as monomers. If they were binding as dimers, the reader should verify that they would appear in the final expressions with a power of two (see Exercise 2.3).

Note also that in this derivation we have assumed that the binding is competitive, that is, B_a and B_r cannot simultaneously bind to A. If they can bind simultaneously to A, we have to include another complex comprising B_a, B_r and A. Denoting this new complex by C', we must add the two additional reactions

$$C_a + B_r \underset{\bar{d}}{\overset{\bar{a}}{\rightleftharpoons}} C', \qquad C_r + B_a \underset{\bar{d}'}{\overset{\bar{a}'}{\rightleftharpoons}} C',$$

and we should modify the conservation law for A to $A_{tot} = A + C_a + C_r + C'$. The reader can verify that in this case a mixed term B_rB_a appears in the equilibrium expressions (see Exercise 2.4).

Enzymatic reaction. A general enzymatic reaction can be written as

$$E + S \underset{d}{\overset{a}{\rightleftharpoons}} C \overset{k}{\rightarrow} E + P,$$

in which E is an enzyme, S is the substrate to which the enzyme binds to form the complex C = ES, and P is the product resulting from the modification of the substrate S due to the binding with the enzyme E. Here, a and d are the association and dissociation rate constants as before, and k is the catalytic rate constant. Enzymatic reactions are very common and include phosphorylation as we have seen in Example 2.1 and as we will see in more detail in the sequel. The corresponding ODE model is given by

$$\frac{dS}{dt} = -aE \cdot S + dC, \qquad \frac{dC}{dt} = aE \cdot S - (d+k)C,$$
$$\frac{dE}{dt} = -aE \cdot S + dC + kC, \qquad \frac{dP}{dt} = kC.$$

The total enzyme concentration is usually constant and denoted by E_{tot}, so that $E + C = E_{\text{tot}}$. Substituting $E = E_{\text{tot}} - C$ in the above equations, we obtain

$$\frac{dS}{dt} = -a(E_{\text{tot}} - C) \cdot S + dC, \qquad \frac{dC}{dt} = a(E_{\text{tot}} - C) \cdot S - (d+k)C,$$
$$\frac{dE}{dt} = -a(E_{\text{tot}} - C) \cdot S + dC + kC, \qquad \frac{dP}{dt} = kC.$$

This system cannot be solved analytically, therefore, assumptions must be used in order to reduce it to a simpler form. Michaelis and Menten assumed that the conversion of E and S to C and vice versa is much faster than the decomposition of C into E and P. Under this assumption and letting the initial concentration $S(0)$ be sufficiently large (see Example 3.12), C immediately reaches its steady state value (while P is still changing). This approximation is called the *quasi-steady state approximation* and the mathematical conditions on the parameters that justify it will be dealt with in Section 3.5.

The steady state value of C is given by solving $a(E_{\text{tot}} - C)S - (d+k)C = 0$ for C, which gives

$$C = \frac{E_{\text{tot}}S}{S + K_m}, \quad \text{with} \quad K_m = \frac{d+k}{a},$$

in which the constant K_m is called the *Michaelis-Menten constant*. Letting $V_{\max} = kE_{\text{tot}}$, the resulting kinetics

$$\frac{dP}{dt} = k\frac{E_{\text{tot}}S}{S + K_m} = V_{\max}\frac{S}{S + K_m} \tag{2.9}$$

are called *Michaelis-Menten kinetics*.

The constant V_{max} is called the maximal velocity (or maximal flux) of modification and it represents the maximal rate that can be obtained when the enzyme is completely saturated by the substrate. The value of K_m corresponds to the value of S that leads to a half-maximal value of the production rate of P. When the enzyme

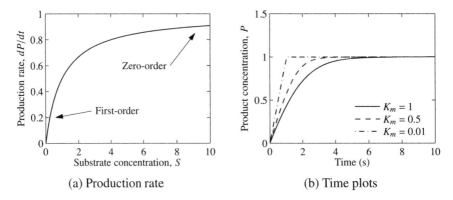

(a) Production rate   (b) Time plots

Figure 2.5: Enzymatic reactions. (a) Transfer curve showing the production rate for P as a function of substrate concentration for $K_m = 1$. (b) Time plots of product $P(t)$ for different values of the K_m. In the plots $S_{tot} = 1$ and $V_{max} = 1$.

complex can be neglected with respect to the total substrate amount S_{tot}, we have that $S_{tot} = S + P + C \approx S + P$, so that the above equation can be also rewritten as

$$\frac{dP}{dt} = \frac{V_{max}(S_{tot} - P)}{(S_{tot} - P) + K_m}.$$

When $K_m \ll S_{tot}$ and the substrate has not yet been all converted to product, that is, $S \gg K_m$, we have that the rate of product formation becomes approximately $dP/dt \approx V_{max}$, which is the maximal speed of reaction. Since this rate is constant and does not depend on the reactant concentrations, it is usually referred to as *zero-order kinetics*. In this case, the system is said to operate in the zero-order regime. If instead $S \ll K_m$, the rate of product formation becomes $dP/dt \approx V_{max}/K_m S$, which is linear with the substrate concentration S. This production rate is referred to as *first-order kinetics* and the system is said to operate in the first-order regime (see Figure 2.5).

2.2   Transcription and translation

In this section we consider the processes of transcription and translation, using the modeling techniques described in the previous section to capture the fundamental dynamic behavior. Models of transcription and translation can be done at a variety of levels of detail and which model to use depends on the questions that one wants to consider. We present several levels of modeling here, starting with a fairly detailed set of reactions describing transcription and translation and ending with highly simplified ordinary differential equation models that can be used when we are only interested in average production rate of mRNA and proteins at relatively long time scales.

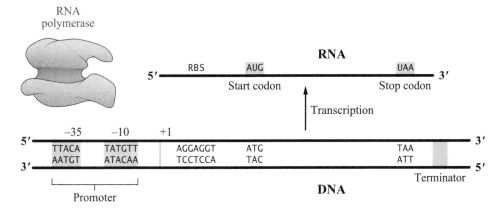

Figure 2.6: Geometric structure of DNA. The layout of the DNA is shown at the top. RNA polymerase binds to the promoter region of the DNA and transcribes the DNA starting at the +1 site and continuing to the termination site. The transcribed mRNA strand has the ribosome binding site (RBS) where the ribosomes bind, the start codon where translation starts and the stop codon where translation ends.

The central dogma: Production of proteins

The genetic material inside a cell, encoded in its DNA, governs the response of a cell to various conditions. DNA is organized into collections of genes, with each gene encoding a corresponding protein that performs a set of functions in the cell. The activation and repression of genes are determined through a series of complex interactions that give rise to a remarkable set of circuits that perform the functions required for life, ranging from basic metabolism to locomotion to procreation. Genetic circuits that occur in nature are robust to external disturbances and can function in a variety of conditions. To understand how these processes occur (and some of the dynamics that govern their behavior), it will be useful to present a relatively detailed description of the underlying biochemistry involved in the production of proteins.

DNA is a double stranded molecule with the "direction" of each strand specified by looking at the geometry of the sugars that make up its backbone. The complementary strands of DNA are composed of a sequence of nucleotides that consist of a sugar molecule (deoxyribose) bound to one of four bases: adenine (A), cytocine (C), guanine (G) and thymine (T). The coding region (by convention the top row of a DNA sequence when it is written in text form) is specified from the 5′ end of the DNA to the 3′ end of the DNA. (The 5′ and 3′ refer to carbon locations on the deoxyribose backbone that are involved in linking together the nucleotides that make up DNA.) The DNA that encodes proteins consists of a promoter region, regulator regions (described in more detail below), a coding region and a termination region (see Figure 2.6). We informally refer to this entire sequence of DNA as a gene.

Expression of a gene begins with the *transcription* of DNA into mRNA by RNA

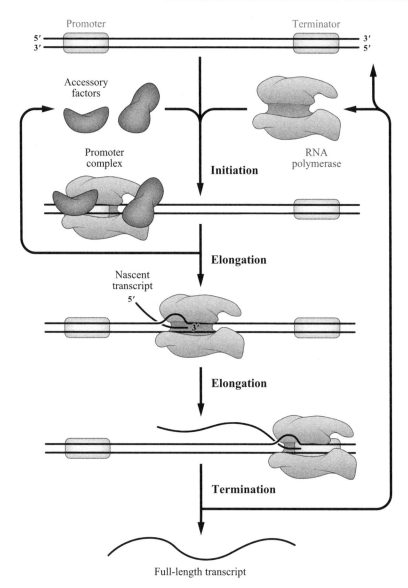

Figure 2.7: Production of messenger RNA from DNA. RNA polymerase, along with other accessory factors, binds to the promoter region of the DNA and then "opens" the DNA to begin transcription (initiation). As RNA polymerase moves down the DNA in the transcription elongation complex (TEC), it produces an RNA transcript (elongation), which is later translated into a protein. The process ends when the RNA polymerase reaches the terminator (termination). Figure adapted from Courey [20].

polymerase, as illustrated in Figure 2.7. RNA polymerase enzymes are present in the nucleus (for eukaryotes) or cytoplasm (for prokaryotes) and must localize and bind to the promoter region of the DNA template. Once bound, the RNA polymerase "opens" the double stranded DNA to expose the nucleotides that make up the sequence. This reaction, called *isomerization*, is said to transform the RNA polymerase and DNA from a *closed complex* to an *open complex*. After the open complex is formed, RNA polymerase begins to travel down the DNA strand and constructs an mRNA sequence that matches the 5′ to 3′ sequence of the DNA to which it is bound. By convention, we number the first base pair that is transcribed as +1 and the base pair prior to that (which is not transcribed) is labeled as -1. The promoter region is often shown with the -10 and -35 regions indicated, since these regions contain the nucleotide sequences to which the RNA polymerase enzyme binds (the locations vary in different cell types, but these two numbers are typically used).

The RNA strand that is produced by RNA polymerase is also a sequence of nucleotides with a sugar backbone. The sugar for RNA is ribose instead of deoxyribose and mRNA typically exists as a single stranded molecule. Another difference is that the base thymine (T) is replaced by uracil (U) in RNA sequences. RNA polymerase produces RNA one base pair at a time, as it moves from in the 5′ to 3′ direction along the DNA coding region. RNA polymerase stops transcribing DNA when it reaches a *termination region* (or *terminator*) on the DNA. This termination region consists of a sequence that causes the RNA polymerase to unbind from the DNA. The sequence is not conserved across species and in many cells the termination sequence is sometimes "leaky," so that transcription will occasionally occur across the terminator.

Once the mRNA is produced, it must be translated into a protein. This process is slightly different in prokaryotes and eukaryotes. In prokaryotes, there is a region of the mRNA in which the ribosome (a molecular complex consisting of both proteins and RNA) binds. This region, called the *ribosome binding site (RBS)*, has some variability between different cell species and between different genes in a given cell. The Shine-Dalgarno sequence, AGGAGG, is the consensus sequence for the RBS. (A consensus sequence is a pattern of nucleotides that implements a given function across multiple organisms; it is not exactly conserved, so some variations in the sequence will be present from one organism to another.)

In eukaryotes, the RNA must undergo several additional steps before it is translated. The RNA sequence that has been created by RNA polymerase consists of *introns* that must be spliced out of the RNA (by a molecular complex called the spliceosome), leaving only the *exons*, which contain the coding region for the protein. The term *pre-mRNA* is often used to distinguish between the raw transcript and the spliced mRNA sequence, which is called *mature mRNA*. In addition to splicing, the mRNA is also modified to contain a *poly(A)* (polyadenine) *tail*, consisting of a long sequence of adenine (A) nucleotides on the 3′ end of the mRNA. This processed sequence is then transported out of the nucleus into the cytoplasm, where the ribosomes can bind to it.

Unlike prokaryotes, eukaryotes do not have a well-defined ribosome binding se-

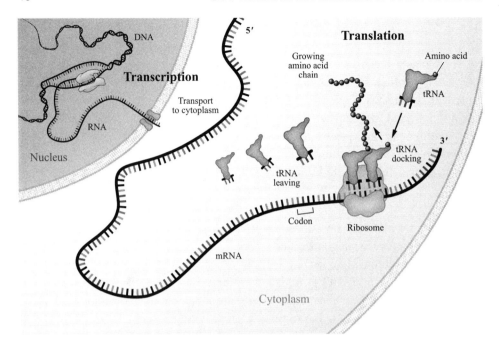

Figure 2.8: Translation is the process of translating the sequence of a messenger RNA (mRNA) molecule to a sequence of amino acids during protein synthesis. The genetic code describes the relationship between the sequence of base pairs in a gene and the corresponding amino acid sequence that it encodes. In the cell cytoplasm, the ribosome reads the sequence of the mRNA in groups of three bases to assemble the protein. Figure and caption courtesy the National Human Genome Research Institute.

quence and hence the process of the binding of the ribosome to the mRNA is more complicated. The *Kozak sequence*, A/GCCACCAUGG, is the rough equivalent of the ribosome binding site, where the underlined AUG is the start codon (described below). However, mRNA lacking the Kozak sequence can also be translated.

Once the ribosome is bound to the mRNA, it begins the process of *translation*. Proteins consist of a sequence of amino acids, with each amino acid specified by a codon that is used by the ribosome in the process of translation. Each codon consists of three base-pairs and corresponds to one of the twenty amino acids or a "stop" codon. The ribosome translates each codon into the corresponding amino acid using transfer RNA (tRNA) to integrate the appropriate amino acid (which binds to the tRNA) into the polypeptide chain, as shown in Figure 2.8. The start codon (AUG) specifies the location at which translation begins, as well as coding for the amino acid methionine (a modified form is used in prokaryotes). All subsequent codons are translated by the ribosome into the corresponding amino acid until it reaches one of the stop codons (typically UAA, UAG and UGA).

The sequence of amino acids produced by the ribosome is a polypeptide chain that folds on itself to form a protein. The process of folding is complicated and

Table 2.1: Rates of core processes involved in the creation of proteins from DNA in *E. coli*.

Process	Characteristic rate	Source
mRNA transcription rate	24-29 bp/s	[13]
Protein translation rate	12–21 aa/s	[13]
Maturation time (fluorescent proteins)	6–60 min	[13]
mRNA half-life	~ 100 s	[103]
E. coli cell division time	20–40 min	[13]
Yeast cell division time	70–140 min	[13]
Protein half-life	$\sim 5 \times 10^4$ s	[103]
Protein diffusion along DNA	up to 10^4 bp/s	[78]
RNA polymerase dissociation constant	~ 0.3–10,000 nM	[13]
Open complex formation kinetic rate	~ 0.02 s^{-1}	[13]
Transcription factor dissociation constant	~ 0.02–10,000 nM	[13]

involves a variety of chemical interactions that are not completely understood. Additional post-translational processing of the protein can also occur at this stage, until a folded and functional protein is produced. It is this molecule that is able to bind to other species in the cell and perform the chemical reactions that underlie the behavior of the organism. The *maturation time* of a protein is the time required for the polypeptide chain to fold into a functional protein.

Each of the processes involved in transcription, translation and folding of the protein takes time and affects the dynamics of the cell. Table 2.1 shows representative rates of some of the key processes involved in the production of proteins. In particular, the dissociation constant of RNA polymerase from the DNA promoter has a wide range of values depending on whether the binding is enhanced by activators (as we will see in the sequel), in which case it can take very low values. Similarly, the dissociation constant of transcription factors with DNA can be very low in the case of specific binding and substantially larger for non-specific binding. It is important to note that each of these steps is highly stochastic, with molecules binding together based on some propensity that depends on the binding energy but also the other molecules present in the cell. In addition, although we have described everything as a sequential process, each of the steps of transcription, translation and folding are happening simultaneously. In fact, there can be multiple RNA polymerases that are bound to the DNA, each producing a transcript. In prokaryotes, as soon as the ribosome binding site has been transcribed, the ribosome can bind and begin translation. It is also possible to have multiple ribosomes bound to a single piece of mRNA. Hence the overall process can be extremely stochastic and asynchronous.

Reaction models

The basic reactions that underlie transcription include the diffusion of RNA polymerase from one part of the cell to the promoter region, binding of an RNA polymerase to the promoter, isomerization from the closed complex to the open complex, and finally the production of mRNA, one base-pair at a time. To capture this set of reactions, we keep track of the various forms of RNA polymerase according to its location and state: $RNAP^c$ represents RNA polymerase in the cytoplasm, $RNAP^p$ represents RNA polymerase in the promoter region, and $RNAP^d$ is RNA polymerase non-specifically bound to DNA. We must similarly keep track of the state of the DNA, to ensure that multiple RNA polymerases do not bind to the same section of DNA. Thus we can write DNA^p for the promoter region, DNA^i for the ith section of the gene of interest and DNA^t for the termination sequence. We write $RNAP : DNA$ to represent RNA polymerase bound to DNA (assumed closed) and $RNAP : DNA^o$ to indicate the open complex. Finally, we must keep track of the mRNA that is produced by transcription: we write $mRNA^i$ to represent an mRNA strand of length i and assume that the length of the gene of interest is N.

Using these various states of the RNA polymerase and locations on the DNA, we can write a set of reactions modeling the basic elements of transcription as

$$
\begin{aligned}
\text{Binding to DNA:} \quad & RNAP^c \rightleftharpoons RNAP^d, \\
\text{Diffusion along DNA:} \quad & RNAP^d \rightleftharpoons RNAP^p, \\
\text{Binding to promoter:} \quad & RNAP^p + DNA^p \rightleftharpoons RNAP : DNA^p, \\
\text{Isomerization:} \quad & RNAP : DNA^p \longrightarrow RNAP : DNA^o, \\
\text{Start of transcription:} \quad & RNAP : DNA^o \longrightarrow RNAP : DNA^1 + DNA^p, \\
\text{mRNA creation:} \quad & RNAP : DNA^1 \longrightarrow RNAP : DNA^2 : mRNA^1, \\
\text{Elongation:} \quad & RNAP : DNA^{i+1} : mRNA^i \\
& \qquad\qquad \longrightarrow RNAP : DNA^{i+2} : mRNA^{i+1}, \\
\text{Binding to terminator:} \quad & RNAP{:}DNA^N : mRNA^{N-1} \\
& \qquad\qquad \longrightarrow RNAP : DNA^t + mRNA^N, \\
\text{Termination:} \quad & RNAP : DNA^t \longrightarrow RNAP^c, \\
\text{Degradation:} \quad & mRNA^N \longrightarrow \emptyset.
\end{aligned}
$$

$$(2.10)$$

Note that at the start of transcription we "release" the promoter region of the DNA, thus allowing a second RNA polymerase to bind to the promoter while the first RNA polymerase is still transcribing the gene. This allows the same DNA strand to be transcribed by multiple RNA polymerase at the same time. The species $RNAP : DNA^{i+1} : mRNA^i$ represents RNA polymerases bound at the $(i+1)$th section of DNA with an elongating mRNA strand of length i attached to it. Upon binding to the terminator region, the RNA polymerase releases the full mRNA strand

$mRNA^N$. This mRNA has the ribosome binding site at which ribosomes can bind to start translation. The main difference between prokaryotes and eukaryotes is that in eukaryotes the RNA polymerase remains in the nucleus and the $mRNA^N$ must be spliced and transported to the cytoplasm before ribosomes can start translation. As a consequence, the start of translation can occur only after $mRNA^N$ has been produced. For simplicity of notation, we assume here that the entire mRNA strand should be produced before ribosomes can start translation. In the procaryotic case, instead, translation can start even for an mRNA strand that is still elongating (see Exercise 2.6).

A similar set of reactions can be written to model the process of translation. Here we must keep track of the binding of the ribosome to the ribosome binding site (RBS) of $mRNA^N$, translation of the mRNA sequence into a polypeptide chain, and folding of the polypeptide chain into a functional protein. Specifically, we must keep track of the various states of the ribosome bound to different codons on the mRNA strand. We thus let Ribo : $mRNA^{RBS}$ denote the ribosome bound to the ribosome binding site of $mRNA^N$, Ribo : $mRNA^{AAi}$ the ribosome bound to the ith codon (corresponding to an amino acid, indicated by the superscript AA), Ribo : $mRNA^{start}$ and Ribo : $mRNA^{stop}$ the ribosome bound to the start and stop codon, respectively. We also let PPC^i denote the polypeptide chain consisting of i amino acids. Here, we assume that the protein of interest has M amino acids. The reactions describing translation can then be written as

$$
\begin{aligned}
\text{Binding to RBS:} \quad & \text{Ribo} + \text{mRNA}^N \rightleftharpoons \text{Ribo} : \text{mRNA}^{RBS}, \\
\text{Start of translation:} \quad & \text{Ribo} : \text{mRNA}^{RBS} \longrightarrow \text{Ribo} : \text{mRNA}^{start} + \text{mRNA}^N, \\
\text{Polypeptide chain creation:} \quad & \text{Ribo} : \text{mRNA}^{start} \longrightarrow \text{Ribo} : \text{mRNA}^{AA2} : \text{PPC}^1, \\
\text{Elongation, } i = 1,\ldots,M: \quad & \text{Ribo} : \text{mRNA}^{AA(i+1)} : \text{PPC}^i \\
& \qquad\qquad \longrightarrow \text{Ribo} : \text{mRNA}^{AA(i+2)} : \text{PPC}^{i+1}, \\
\text{Stop codon:} \quad & \text{Ribo} : \text{mRNA}^{AAM} : \text{PPC}^{M-1} \\
& \qquad\qquad \longrightarrow \text{Ribo} : \text{mRNA}^{stop} + \text{PPC}^M, \\
\text{Release of mRNA:} \quad & \text{Ribo} : \text{mRNA}^{stop} \longrightarrow \text{Ribo}, \\
\text{Folding:} \quad & \text{PPC}^M \longrightarrow \text{protein}, \\
\text{Degradation:} \quad & \text{protein} \longrightarrow \emptyset.
\end{aligned}
$$

(2.11)

As in the case of transcription, we see that these reactions allow multiple ribosomes to translate the same piece of mRNA by freeing up $mRNA^N$. After M amino acids have been chained together, the M-long polypeptide chain PPC^M is released, which then folds into a protein. As complex as these reactions are, they do not directly capture a number of physical phenomena such as ribosome queuing, wherein ribosomes cannot pass other ribosomes that are ahead of them on the mRNA chain. Additionally, we have not accounted for the existence and effects of the 5′ and

3′ untranslated regions (UTRs) of a gene and we have also left out various error correction mechanisms in which ribosomes can step back and release an incorrect amino acid that has been incorporated into the polypeptide chain. We have also left out the many chemical species that must be present in order for a variety of the reactions to happen (NTPs for mRNA production, amino acids for protein production, etc.). Incorporation of these effects requires additional reactions that track the many possible states of the molecular machinery that underlies transcription and translation. For more detailed models of translation, the reader is referred to [3].

When the details of the isomerization, start of transcription (translation), elongation, and termination are not relevant for the phenomenon to be studied, the transcription and translation reactions are lumped into much simpler reduced reactions. For transcription, these reduced reactions take the form:

$$\begin{aligned} \text{RNAP} + \text{DNA}^p &\rightleftharpoons \text{RNAP:DNA}^p, \\ \text{RNAP:DNA}^p &\longrightarrow \text{mRNA} + \text{RNAP} + \text{DNA}^p, \\ \text{mRNA} &\longrightarrow \emptyset, \end{aligned} \tag{2.12}$$

in which the second reaction lumps together isomerization, start of transcription, elongation, mRNA creation, and termination. Similarly, for the translation process, the reduced reactions take the form:

$$\begin{aligned} \text{Ribo} + \text{mRNA} &\rightleftharpoons \text{Ribo:mRNA}, \\ \text{Ribo:mRNA} &\longrightarrow \text{protein} + \text{mRNA} + \text{Ribo}, \\ \text{Ribo:mRNA} &\longrightarrow \text{Ribo}, \\ \text{protein} &\longrightarrow \emptyset, \end{aligned} \tag{2.13}$$

in which the second reaction lumps the start of translation, elongation, folding, and termination. The third reaction models the fact that mRNA can also be degraded when bound to ribosomes when the ribosome binding site is left free. The process of mRNA degradation occurs through RNase enzymes binding to the ribosome binding site and cleaving the mRNA strand. It is known that the ribosome binding site cannot be both bound to the ribosome and to the RNase [68]. However, the species Ribo : mRNA is a lumped species encompassing configurations in which ribosomes are bound on the mRNA strand but not on the ribosome binding site. Hence, we also allow this species to be degraded by RNase.

Reaction rate equations

Given a set of reactions, the various stochastic processes that underlie detailed models of transcription and translation can be specified using the stochastic modeling framework described briefly in the previous section. In particular, using either models of binding energy or measured rates, we can construct propensity functions for each of the many reactions that lead to production of proteins, including the motion of RNA polymerase and the ribosome along DNA and RNA. For many problems in which the detailed stochastic nature of the molecular dynamics of the cell are important, these models are the most relevant and they are covered in some detail in Chapter 4.

Alternatively, we can move to the reaction rate formalism and model the reactions using differential equations. To do so, we must compute the various reaction rates, which can be obtained from the propensity functions or measured experimentally. In moving to this formalism, we approximate the concentrations of various species as real numbers (though this may not be accurate for some species that exist at low molecular counts in the cell). Despite these approximations, in many situations the reaction rate equations are sufficient, particularly if we are interested in the average behavior of a large number of cells.

In some situations, an even simpler model of the transcription, translation and folding processes can be utilized. Let the "active" mRNA be the mRNA that is available for translation by the ribosome. We model its concentration through a simple time delay of length τ^m that accounts for the transcription of the ribosome binding site in prokaryotes or splicing and transport from the nucleus in eukaryotes. If we assume that RNA polymerase binds to DNA at some average rate (which includes both the binding and isomerization reactions) and that transcription takes some fixed time (depending on the length of the gene), then the process of transcription can be described using the delay differential equation

$$\frac{dm_P}{dt} = \alpha - \mu m_P - \bar{\delta} m_P, \qquad m_P^*(t) = e^{-\mu \tau^m} m_P(t - \tau^m), \qquad (2.14)$$

where m_P is the concentration of mRNA for protein P, m_P^* is the concentration of active mRNA, α is the rate of production of the mRNA for protein P, μ is the growth rate of the cell (which results in dilution of the concentration) and $\bar{\delta}$ is the rate of degradation of the mRNA. Since the dilution and degradation terms are of the same form, we will often combine these terms in the mRNA dynamics and use a single coefficient $\delta = \mu + \bar{\delta}$. The exponential factor in the second expression in equation (2.14) accounts for dilution due to the change in volume of the cell, where μ is the cell growth rate. The constants α and δ capture the average rates of production and decay, which in turn depend on the more detailed biochemical reactions that underlie transcription.

Once the active mRNA is produced, the process of translation can be described via a similar ordinary differential equation that describes the production of a functional protein:

$$\frac{dP}{dt} = \kappa m_P^* - \gamma P, \qquad P^f(t) = e^{-\mu \tau^f} P(t - \tau^f). \qquad (2.15)$$

Here P represents the concentration of the polypeptide chain for the protein, and P^f represents the concentration of functional protein (after folding). The parameters that govern the dynamics are κ, the rate of translation of mRNA; γ, the rate of degradation and dilution of P; and τ^f, the time delay associated with folding and other processes required to make the protein functional. The exponential term again accounts for dilution due to cell growth. The degradation and dilution term, parameterized by γ, captures both the rate at which the polypeptide chain is degraded and the rate at which the concentration is diluted due to cell growth.

It will often be convenient to write the dynamics for transcription and translation in terms of the functional mRNA and functional protein. Differentiating the expression for m_P^*, we see that

$$
\begin{aligned}
\frac{dm_P^*(t)}{dt} &= e^{-\mu\tau^m}\frac{dm_P}{dt}(t-\tau^m) \\
&= e^{-\mu\tau^m}(\alpha - \delta m_P(t-\tau^m)) = \bar\alpha - \delta m_P^*(t),
\end{aligned}
\tag{2.16}
$$

where $\bar\alpha = e^{-\mu\tau^m}\alpha$. A similar expansion for the active protein dynamics yields

$$
\frac{dP^f(t)}{dt} = \bar\kappa m_P^*(t-\tau^f) - \gamma P^f(t),
\tag{2.17}
$$

where $\bar\kappa = e^{-\mu\tau^f}\kappa$. We shall typically use equations (2.16) and (2.17) as our (reduced) description of protein folding, dropping the superscript f and overbars when there is no risk of confusion. Also, in the presence of different proteins, we will attach subscripts to the parameters to denote the protein to which they refer.

In many situations the time delays described in the dynamics of protein production are small compared with the time scales at which the protein concentration changes (depending on the values of the other parameters in the system). In such cases, we can simplify our model of the dynamics of protein production even further and write

$$
\frac{dm_P}{dt} = \alpha - \delta m_P, \qquad \frac{dP}{dt} = \kappa m_P - \gamma P.
\tag{2.18}
$$

Note that we here have dropped the superscripts $*$ and f since we are assuming that all mRNA is active and proteins are functional and dropped the overbar on α and κ since we are assuming the time delays are negligible. The value of α increases with the strength of the promoter while the value of κ increases with the strength of the ribosome binding site. These strengths, in turn, can be affected by changing the specific base-pair sequences that constitute the promoter RNA polymerase binding region and the ribosome binding site.

Finally, the simplest model for protein production is one in which we only keep track of the basal rate of production of the protein, without including the mRNA dynamics. This essentially amounts to assuming the mRNA dynamics reach steady state quickly and replacing the first differential equation in (2.18) with its equilibrium value. This is often a good assumption as mRNA degration is usually about 100 times faster than protein degradation (see Table 2.1). Thus we obtain

$$
\frac{dP}{dt} = \beta - \gamma P, \qquad \beta := \kappa\frac{\alpha}{\delta}.
$$

This model represents a simple first-order, linear differential equation for the rate of production of a protein. In many cases this will be a sufficiently good approximate model, although we will see that in some cases it is too simple to capture the observed behavior of a biological circuit.

2.3 Transcriptional regulation

The operation of a cell is governed in part by the selective expression of genes in the DNA of the organism, which control the various functions the cell is able to perform at any given time. Regulation of protein activity is a major component of the molecular activities in a cell. By turning genes on and off, and modulating their activity in more fine-grained ways, the cell controls its many metabolic pathways, responds to external stimuli, differentiates into different cell types as it divides, and maintains the internal state of the cell required to sustain life.

The regulation of gene expression and protein activity is accomplished through a variety of molecular mechanisms, as discussed in Section 1.2 and illustrated in Figure 2.9. At each stage of the processing from a gene to a protein, there are potential mechanisms for regulating the production processes. The remainder of this section will focus on transcriptional control and the next section on selected mechanisms for controlling protein activity. We will focus on prokaryotic mechanisms.

Transcriptional regulation of protein production

The simplest forms of transcriptional regulation are repression and activation, both controlled through proteins called *transcription factors*. In the case of *repression*, the presence of a transcription factor (often a protein that binds near the promoter) turns off the transcription of the gene and this type of regulation is often called negative regulation or "down regulation." In the case of *activation* (or positive regulation), transcription is enhanced when an activator protein binds to the promoter site (facilitating binding of the RNA polymerase).

Repression. A common mechanism for repression is that a protein binds to a region of DNA near the promoter and blocks RNA polymerase from binding. The region of DNA to which the repressor protein binds is called an *operator region* (see Figure 2.10a). If the operator region overlaps the promoter, then the presence of a protein at the promoter can "block" the DNA at that location and transcription cannot initiate. Repressor proteins often bind to DNA as dimers or pairs of dimers (effectively tetramers). Figure 2.10b shows some examples of repressors bound to DNA.

A related mechanism for repression is DNA looping. In this setting, two repressor complexes (often dimers) bind in different locations on the DNA and then bind to each other. This can create a loop in the DNA and block the ability of RNA polymerase to bind to the promoter, thus inhibiting transcription. Figure 2.11 shows an

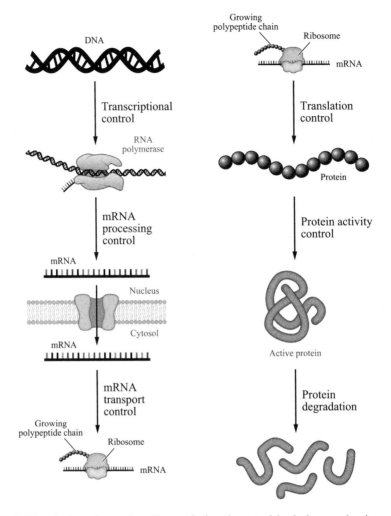

Figure 2.9: Regulation of proteins. Transcriptional control includes mechanisms to tune the rate at which mRNA is produced from DNA, while translation control includes mechanisms to tune the rate at which the protein polypeptide chain is produced from mRNA. Protein activity control encompasses many processes, such as phosphorylation, methylation, and allosteric modification. Figure adapted from Phillips, Kondev and Theriot [78].

example of this type of repression, in the *lac* operon. (An *operon* is a set of genes that is under control of a single promoter.)

Activation. The process of activation of a gene requires that an activator protein be present in order for transcription to occur. In this case, the protein must work to either recruit or enable RNA polymerase to begin transcription.

The simplest form of activation involves a protein binding to the DNA near the promoter in such a way that the combination of the activator and the promoter sequence bind RNA polymerase. Figure 2.12 illustrates the basic concept.

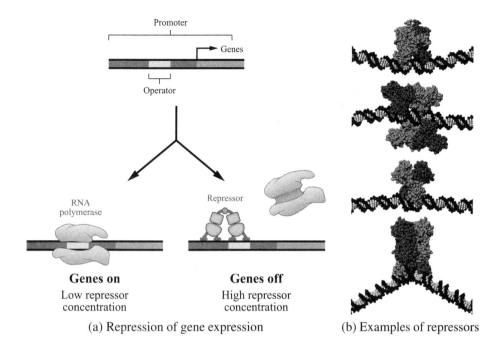

(a) Repression of gene expression   (b) Examples of repressors

Figure 2.10: Repression of gene expression. A repressor protein binds to operator sites on the gene promoter and blocks the binding of RNA polymerase to the promoter, so that the gene is off. Figures adapted from Phillips, Kondev and Theriot [78]. Copyright 2009 from *Physical Biology of the Cell* by Phillips et al. Reproduced by permission of Garland Science/Taylor & Francis LLC.

Another mechanism for activation of transcription, specific to prokaryotes, is the use of *sigma factors*. Sigma factors are part of a modular set of proteins that bind to RNA polymerase and form the molecular complex that performs transcription. Different sigma factors enable RNA polymerase to bind to different promoters, so the sigma factor acts as a type of activating signal for transcription. Table 2.2 lists some of the common sigma factors in bacteria. One of the uses of sigma factors is to produce certain proteins only under special conditions, such as when the

Table 2.2: Sigma factors in *E. coli* [2].

Sigma factor	Promoters recognized
σ^{70}	most genes
σ^{32}	genes associated with heat shock
σ^{38}	genes involved in stationary phase and stress response
σ^{28}	genes involved in motility and chemotaxis
σ^{24}	genes dealing with misfolded proteins in the periplasm

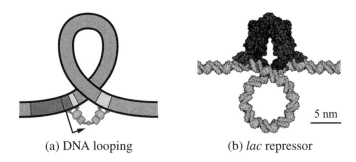

(a) DNA looping   (b) *lac* repressor

Figure 2.11: Repression via DNA looping. A repressor protein can bind simultaneously to two DNA sites downstream of the start of transcription, thus creating a loop that prevents RNA polymerase from transcribing the gene. Figures adapted from Phillips, Kondev and Theriot [78]. Copyright 2009 from *Physical Biology of the Cell* by Phillips et al. Reproduced by permission of Garland Science/Taylor & Francis LLC.

cell undergoes *heat shock*. Another use is to control the timing of the expression of certain genes, as illustrated in Figure 2.13.

Inducers. A feature that is present in some types of transcription factors is the existence of an inducer molecule that combines with the protein to either activate or inactivate its function. A *positive inducer* is a molecule that must be present in order for repression or activation to occur. A *negative inducer* is one in which the presence of the inducer molecule blocks repression or activation, either by changing the shape of the transcription factor protein or by blocking active sites on the protein that would normally bind to the DNA. Figure 2.14 summarizes the various possibilities. Common examples of repressor-inducer pairs include *lacI* and lactose (or IPTG), and *tetR* and aTc. Lactose/IPTG and aTc are both negative inducers, so their presence causes the otherwise repressed gene to be expressed. An example of a positive inducer is cyclic AMP (cAMP), which acts as a positive inducer for the CAP activator.

Combinatorial promoters. In addition to promoters that can take either a repressor or an activator as the sole input transcription factor, there are combinatorial promoters that can take both repressors and activators as input transcription factors. This allows genes to be switched on and off based on more complex conditions, represented by the concentrations of two or more activators or repressors.

Figure 2.15 shows one of the classic examples, a promoter for the *lac* system. In the *lac* system, the expression of genes for metabolizing lactose are under the control of a single (combinatorial) promoter. CAP, which is positively induced by cAMP, acts as an activator and LacI (also called "Lac repressor"), which is negatively induced by lactose, acts as a repressor. In addition, the inducer cAMP is expressed only when glucose levels are low. The resulting behavior is that the proteins for metabolizing lactose are expressed only in conditions where there is no glucose (so CAP is active) *and* lactose is present.

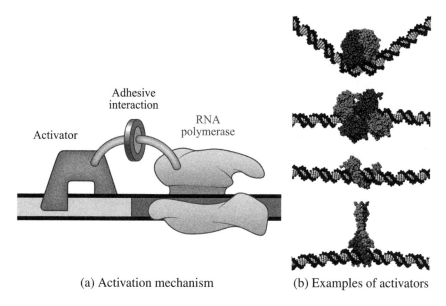

(a) Activation mechanism   (b) Examples of activators

Figure 2.12: Activation of gene expression. (a) Conceptual operation of an activator. The activator binds to DNA upstream of the gene and attracts RNA polymerase to the DNA strand. (b) Examples of activators: catabolite activator protein (CAP), p53 tumor suppressor, zinc finger DNA binding domain and leucine zipper DAN binding domain. Figures adapted from Phillips, Kondev and Theriot [78]. Copyright 2009 from *Physical Biology of the Cell* by Phillips et al. Reproduced by permission of Garland Science/Taylor & Francis LLC.

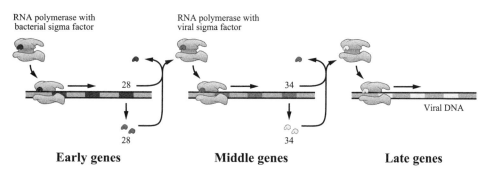

Figure 2.13: Use of sigma factors to control the timing of gene expression in a bacterial virus. Early genes are transcribed by RNA polymerase bound to bacterial sigma factors. One of the early genes, called 28, encodes a sigma-like factor that binds to RNA polymerase and allow it to transcribe middle genes, which in turn produce another sigma-like factor that allows RNA polymerase to transcribe late genes. These late genes produce proteins that form a coat for the viral DNA and lyse the cell. Figure adapted from Alberts et al. [2].

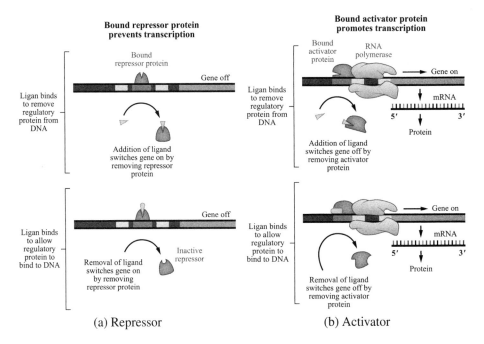

Figure 2.14: Effects of inducers. (a) In the case of repressors, a negative inducer binds to the repressor making it unbind DNA, thus enabling transcription. A positive inducer, by contrast, activates the repressor allowing it to bind DNA. (b) In the case of activators, a negative inducer binds to the activator making it unbind DNA, thus preventing transcription. A positive inducer instead enables the activator to bind DNA, allowing transcription. Figures adapted from Alberts et al. [2].

More complicated combinatorial promoters can also be used to control transcription in two different directions, an example that is found in some viruses.

Antitermination. A final method of activation in prokaryotes is the use of *antitermination*. The basic mechanism involves a protein that binds to DNA and deactivates a site that would normally serve as a termination site for RNA polymerase. Additional genes are located downstream from the termination site, but without a promoter region. Thus, in the absence of the antiterminator protein, these genes are not expressed (or expressed with low probability). However, when the antitermination protein is present, the RNA polymerase maintains (or regains) its contact with the DNA and expression of the downstream genes is enhanced. In this way, antitermination allows downstream genes to be regulated by repressing "premature" termination. An example of an antitermination protein is the protein N in phage λ, which binds to a region of DNA labeled nut (for N utilization), as shown in Figure 2.16 [37].

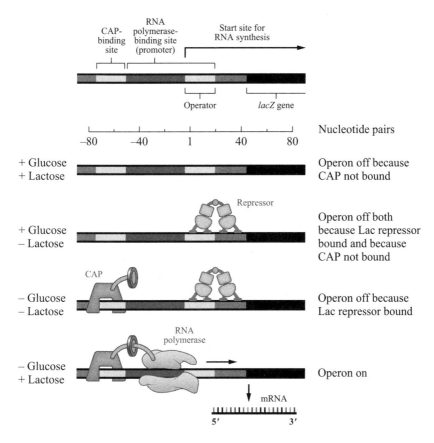

Figure 2.15: Combinatorial logic for the *lac* operator. The CAP-binding site and the operator in the promoter can be both bound by CAP (activator) and by LacI (Lac repressor), respectively. The only configuration in which RNA polymerase can bind the promoter and start transcription is where CAP is bound but LacI is not bound. Figure adapted from Phillips, Kondev and Theriot [78].

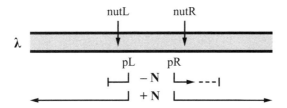

Figure 2.16: Antitermination. Protein N binds to DNA regions labeled nut, enabling transcription of longer DNA sequences. Figure adapted from [37].

Reaction models

We can capture the molecular interactions responsible for transcriptional regulation by modifying the RNA polymerase binding reactions in equation (2.10) to include the binding of the repressor or activator to the promoter. For a repressor (Rep), we have to add a reaction that represents the repressor bound to the promoter DNA^P:

$$\text{Repressor binding:} \quad DNA^P + Rep \rightleftharpoons DNA:Rep.$$

This reaction acts to "sequester" the DNA promoter site so that it is no longer available for binding by RNA polymerase. The strength of the repressor is reflected in the reaction rate constants for the repressor binding reaction. Sometimes, the RNA polymerase can bind to the promoter even when the repressor is bound, usually with lower association rate constant. In this case, the repressor still allows some transcription even when bound to the promoter and the repressor is said to be "leaky."

The modifications for an activator (Act) are a bit more complicated, since we have to modify the reactions to require the presence of the activator before RNA polymerase can bind the promoter. One possible mechanism, known as the *recruitment model*, is given by

$$
\begin{aligned}
\text{Activator binding:} \quad & DNA^P + Act \rightleftharpoons DNA^P:Act, \\
\text{RNAP binding w/ activator:} \quad & RNAP^p + DNA^P:Act \rightleftharpoons RNAP:DNA^P:Act, \\
\text{Isomerization:} \quad & RNAP:DNA^P:Act \longrightarrow RNAP:DNA^o:Act, \\
\text{Start of transcription:} \quad & RNAP:DNA^o:Act \longrightarrow RNAP:DNA^1 + DNA^P:Act.
\end{aligned}
$$
$$(2.19)$$

In this model, RNA polymerase cannot bind to the promoter unless the activator is already bound to it. More generally, one can model both the enhanced binding of the RNA polymerase to the promoter in the presence of the activator, as well as the possibility of binding without an activator. This translates into the additional reaction $RNAP^p + DNA^P \rightleftharpoons RNAP:DNA^P$. The relative reaction rates determine how strong the activator is and the "leakiness" of transcription in the absence of the activator. A different model of activation, called allosteric activation, is one in which the RNA polymerase binding rate to DNA is not enhanced by the presence of the activator bound to the promoter, but the open complex (and hence start of transcription) formation can occur only (is enhanced) in the presence of the activator.

A simplified ordinary differential equation model of transcription in the presence of activators or repressors can be obtained by accounting for the fact that transcription factors and RNAP bind to the DNA rapidly when compared to other reactions, such as isomerization and elongation. As a consequence, we can make use of the reduced-order models that describe the quasi-steady state concentrations of proteins bound to DNA as described in Section 2.1. We can consider the competitive binding case to model a strong repressor that prevents RNA polymerase from binding to the DNA. In the sequel, we remove the superscripts "p" and "d" from RNA polymerase to simplify notation. The quasi-steady state concentration of the complex of DNA promoter bound to the repressor will have the expression

$$[\text{DNA}^{\text{P}}{:}\text{Rep}] = \frac{[\text{DNA}]_{\text{tot}}([\text{Rep}]/K_{\text{d}})}{1 + [\text{Rep}]/K_{\text{d}} + [\text{RNAP}]/K'_{\text{d}}}$$

and the steady state amount of DNA promoter bound to the RNA polymerase will be given by

$$[\text{RNAP}{:}\text{DNA}^{\text{P}}] = \frac{[\text{DNA}]_{\text{tot}}([\text{RNAP}]/K'_{\text{d}})}{1 + [\text{RNAP}]/K'_{\text{d}} + [\text{Rep}]/K_{\text{d}}},$$

in which K'_{d} is the dissociation constant of RNA polymerase from the promoter, while K_{d} is the dissociation constant of Rep from the promoter, and $[\text{DNA}]_{\text{tot}}$ represents the total concentration of DNA. The free DNA promoter with RNA polymerase bound will allow transcription, while the complex $\text{DNA}^{\text{P}} : \text{Rep}$ will not allow transcription as it is not bound to RNA polymerase. Using the lumped reactions (2.12) with reaction rate constant k_{f}, this can be modeled as

$$\frac{d[\text{mRNA}]}{dt} = F([\text{Rep}]) - \delta[\text{mRNA}],$$

in which the production rate is given by

$$F([\text{Rep}]) = k_{\text{f}} \frac{[\text{DNA}]_{\text{tot}}\,([\text{RNAP}]/K'_{\text{d}})}{1 + [\text{RNAP}]/K'_{\text{d}} + [\text{Rep}]/K_{\text{d}}}.$$

If the repressor binds to the promoter with cooperativity n, the above expression becomes (see Section 2.1)

$$F([\text{Rep}]) = k_{\text{f}} \frac{[\text{DNA}]_{\text{tot}}\,([\text{RNAP}]/K'_{\text{d}})}{1 + [\text{RNAP}]/K'_{\text{d}} + [\text{Rep}]^n/(K_{\text{m}}K_{\text{d}})},$$

in which K_{m} is the dissociation constant of the reaction of n molecules of Rep binding together. The function F is usually represented in the standard Hill function form

$$F([\text{Rep}]) = \frac{\alpha}{1 + ([\text{Rep}]/K)^n},$$

in which α and K are given by

$$\alpha = \frac{k_{\text{f}}[\text{DNA}]_{\text{tot}}([\text{RNAP}]/K'_{\text{d}})}{1 + ([\text{RNAP}]/K'_{\text{d}})}, \qquad K = \left(K_{\text{m}}K_{\text{d}}(1 + ([\text{RNAP}]/K'_{\text{d}}))\right)^{1/n}.$$

Finally, if the repressor allows RNA polymerase to still bind to the promoter at a small rate (leaky repressor), the above expression can be modified to take the form

$$F([\text{Rep}]) = \frac{\alpha}{1 + ([\text{Rep}]/K)^n} + \alpha_0, \tag{2.20}$$

in which α_0 is the basal expression rate when the promoter is fully repressed, usually referred to as leakiness (see Exercise 2.8).

To model the production rate of mRNA in the case in which an activator Act is required for transcription, we can consider the case in which RNA polymerase binds only when the activator is already bound to the promoter (recruitment model). To simplify the mathematical derivation, we rewrite the reactions (2.19) involving the activator with the lumped transcription reaction (2.12) into the following:

$$
\begin{aligned}
&\text{DNA}^{\text{p}} + \text{Act} \rightleftharpoons \text{DNA}^{\text{p}}{:}\text{Act}, \\
&\text{RNAP} + \text{DNA}^{\text{p}}{:}\text{Act} \rightleftharpoons \text{RNAP}{:}\text{DNA}^{\text{p}}{:}\text{Act}, \\
&\text{RNAP}{:}\text{DNA}^{\text{p}}{:}\text{Act} \xrightarrow{k_{\text{f}}} \text{mRNA} + \text{RNAP} + \text{DNA}^{\text{p}}{:}\text{Act},
\end{aligned}
\tag{2.21}
$$

in which the third reaction lumps together isomerization, start of transcription, elongation and termination. The first and second reactions fit the structure of the cooperative binding model illustrated in Section 2.1. Also, since the third reaction is much slower than the first two, the complex $\text{RNAP}:\text{DNA}^{\text{p}}:\text{Act}$ concentration can be well approximated at its quasi-steady state value. The expression of the quasi-steady state concentration was given in Section 2.1 in correspondence to the cooperative binding model and takes the form

$$
[\text{RNAP}{:}\text{DNA}^{\text{p}}{:}\text{Act}] = \frac{[\text{DNA}]_{\text{tot}}([\text{RNAP}]/K_{\text{d}}')([\text{Act}])/K_{\text{d}}}{1 + ([\text{Act}]/K_{\text{d}})(1 + [\text{RNAP}]/K_{\text{d}}')},
$$

in which K_{d}' is the dissociation constant of RNA polymerase with the complex of DNA bound to Act and K_{d} is the dissociation constant of Act with DNA. When the activator Act binds to the promoter with cooperativity n, the above expression becomes

$$
[\text{RNAP}{:}\text{DNA}^{\text{p}}{:}\text{Act}] = \frac{[\text{DNA}]_{\text{tot}}([\text{RNAP}][\text{Act}]^n)/(K_{\text{d}}K_{\text{d}}'K_{\text{m}})}{1 + ([\text{Act}]^n/K_{\text{d}}K_{\text{m}})(1 + [\text{RNAP}]/K_{\text{d}}')},
$$

in which K_{m} is the dissociation constant of the reaction of n molecules of Act binding together.

In order to write the differential equation for the mRNA concentration, we consider the third reaction in (2.21) along with the above quasi-steady state expressions of $[\text{RNAP}:\text{DNA}^{\text{p}}:\text{Act}]$ to obtain

$$
\frac{d\,[\text{mRNA}]}{dt} = F([\text{Act}]) - \delta[\text{mRNA}],
$$

in which

$$
F([\text{Act}]) = k_{\text{f}} \frac{[\text{DNA}]_{\text{tot}}([\text{RNAP}][\text{Act}]^n)/(K_{\text{d}}K_{\text{d}}'K_{\text{m}})}{1 + ([\text{Act}]^n/K_{\text{d}}K_{\text{m}})(1 + [\text{RNAP}]/K_{\text{d}}')} =: \frac{\alpha([\text{Act}]/K)^n}{1 + ([\text{Act}]/K)^n},
$$

where α and K are implicitly defined. The right-hand side of this expression is in standard Hill function form.

If we assume that RNA polymerase can still bind to DNA even when the activator is not bound, we have an additional basal expression rate α_0 so that the new form of the production rate is given by (see Exercise 2.9)

$$F([\text{Act}]) = \frac{\alpha([\text{Act}]/K)^n}{1 + ([\text{Act}]/K)^n} + \alpha_0. \tag{2.22}$$

As indicated earlier, many activators and repressors operate in the presence of inducers. To incorporate these dynamics in our description, we simply have to add the reactions that correspond to the interaction of the inducer with the relevant protein. For a negative inducer, we can add a reaction in which the inducer binds the regulator protein and effectively sequesters it so that it cannot interact with the DNA. For example, a negative inducer operating on a repressor could be modeled by adding the reaction

$$\text{Rep} + \text{Ind} \rightleftharpoons \text{Rep:Ind}.$$

Since the above reactions are very fast compared to transcription, they can be assumed at the quasi-steady state. Hence, the free amount of repressor that can still bind to the promoter can be calculated by writing the ODE model corresponding to the above reactions and by setting the time derivatives to zero. This yields

$$[\text{Rep}] = \frac{[\text{Rep}]_{\text{tot}}}{1 + [\text{Ind}]/\bar{K}_{\text{d}}},$$

in which $[\text{Rep}]_{\text{tot}} = [\text{Rep}] + [\text{Rep:Ind}]$ is the total amount of repressor (bound and unbound to the inducer) and $\bar{K}_{\text{d}}$ is the dissociation constant of Ind binding to Rep. This expression of the repressor concentration needs to be substituted in the expression of the production rate $F([\text{Rep}])$.

Positive inducers can be handled similarly, except now we have to modify the binding reactions to only work in the presence of a regulatory protein bound to an inducer. For example, a positive inducer on an activator would have the modified reactions

$$\begin{aligned}
\text{Inducer binding:} \quad & \text{Act} + \text{Ind} \rightleftharpoons \text{Act:Ind}, \\
\text{Activator binding:} \quad & \text{DNA}^\text{p} + \text{Act:Ind} \rightleftharpoons \text{DNA}^\text{p}\text{:Act:Ind}, \\
\text{RNAP binding w/ activator:} \quad & \text{RNAP} + \text{DNA}^\text{p}\text{:Act:Ind} \rightleftharpoons \text{RNAP:DNA}^\text{p}\text{:Act:Ind}, \\
\text{Isomerization:} \quad & \text{RNAP:DNA}^\text{p}\text{:Act:Ind} \longrightarrow \text{RNAP:DNA}^\text{o}\text{:Act:Ind}, \\
\text{Start of transcription:} \quad & \text{RNAP:DNA}^\text{o}\text{:Act:Ind} \longrightarrow \text{RNAP:DNA}^1 \\
& \qquad\qquad\qquad\qquad\qquad\quad + \text{DNA}^\text{p}\text{:Act:Ind}.
\end{aligned}$$

Hence, in the expression of the production rate $F([\text{Act}])$, we should substitute the concentration $[\text{Act:Ind}]$ in place of $[\text{Act}]$. This concentration, in turn, is well approximated by its quasi-steady state value since binding reactions are much faster

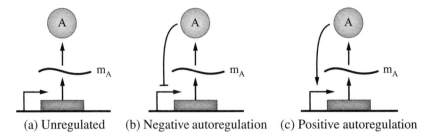

(a) Unregulated   (b) Negative autoregulation   (c) Positive autoregulation

Figure 2.17: Autoregulation of gene expression. In (a) the circuit is unregulated, while (b) shows negative autoregulation and (c) shows positive autoregulation.

than isomerization and transcription, and can be obtained as in the negative inducer case.

Example 2.2 (Autoregulation of gene expression). Consider the circuits shown in Figure 2.17, representing an unregulated gene, a negatively autoregulated gene and a positively autoregulated gene. We want to model the dynamics of the protein A starting from zero initial conditions for the three different cases to understand how the three different circuit topologies affect dynamics.

The dynamics of the three circuits can be written in a common form,

$$\frac{dm_A}{dt} = F(A) - \delta m_A, \qquad \frac{dA}{dt} = \kappa m_A - \gamma A, \qquad (2.23)$$

where $F(A)$ is in one of the following forms:

$$F_{\text{unreg}}(A) = \alpha_B, \qquad F_{\text{repress}}(A) = \frac{\alpha_B}{1 + (A/K)^n} + \alpha_0, \qquad F_{\text{act}}(A) = \frac{\alpha_A (A/K)^n}{1 + (A/K)^n} + \alpha_B.$$

We choose the parameters to be

$$\alpha_A = 0.375 \text{ nM/s}, \qquad \alpha_B = 0.5 \text{ nM/s}, \qquad \alpha_0 = 5 \times 10^{-4} \text{ nM/s},$$
$$\kappa = 0.116 \text{ s}^{-1}, \qquad \delta = 5.78 \times 10^{-3} \text{ s}^{-1}, \qquad \gamma = 1.16 \times 10^{-3} \text{ s}^{-1},$$
$$K = 10^4 \text{ nM}, \qquad n = 2,$$

corresponding to biologically plausible values. Note that the parameters are chosen so that $F(0) \approx \alpha_B$ for each circuit.

Figure 2.18a shows the results of simulations comparing the response of the three circuits. We see that initial increase in protein concentration is identical for each circuit, consistent with our choice of Hill functions and parameters. As the expression level increases, the effects of positive and negative regulation are seen, leading to different steady state expression levels. In particular, the negative feedback circuit reaches a lower steady state expression level while the positive feedback circuit settles to a higher value.

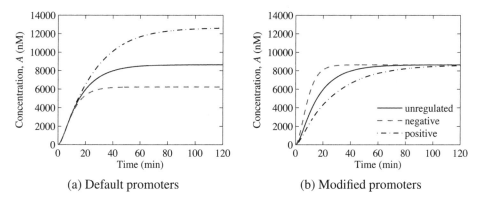

(a) Default promoters   (b) Modified promoters

Figure 2.18: Simulations for autoregulated gene expression. (a) Non-adjusted expression levels. (b) Equalized expression levels.

In some situations, it makes sense to ask whether different circuit topologies have different properties that might lead us to choose one over another. In the case where the circuit is going to be used as part of a more complex pathway, it may make the most sense to compare circuits that produce the same steady state concentration of the protein A. To do this, we must modify the parameters of the individual circuits, which can be done in a number of different ways: we can modify the promoter strengths, degradation rates, or other molecular mechanisms reflected in the parameters.

The steady state expression level for the negative autoregulation case can be adjusted by using a stronger promoter (modeled by α_B) or ribosome binding site (modeled by κ). The equilibrium point for the negative autoregulation case is given by the solution of the equations

$$m_{\mathrm{A},e} = \frac{\alpha K^n}{\delta(K^n + A_e^n)}, \qquad A_e = \frac{\kappa}{\gamma} m_{\mathrm{A},e}.$$

These coupled equations can be solved for $m_{\mathrm{A},e}$ and A_e, but in this case we simply need to find values α_B' and κ' that give the same values as the unregulated case. For example, if we equate the mRNA levels of the unregulated system with that of the negatively autoregulated system, we have

$$\frac{\alpha_B}{\delta} = \frac{1}{\delta}\left(\frac{\alpha_B' K^n}{K^n + A_e^n} + \alpha_0\right) \quad \Longrightarrow \quad \alpha_B' = (\alpha_B - \alpha_0)\frac{K^n + A_e^n}{K^n}, \quad A_e = \frac{\alpha_B \kappa}{\delta \gamma},$$

where A_e is the desired equilibrium value (which we choose using the unregulated case as a guide).

A similar calculation can be done for the case of positive autoregulation, in this case decreasing the promoter parameters α_A and α_B so that the steady state values match. A simple way to do this is to leave α_A unchanged and decrease α_B

to account for the positive feedback. Solving for α'_B to give the same mRNA levels as the unregulated case yields

$$\alpha'_B = \alpha_B - \alpha_A \frac{A_e^n}{K^n + A_e^n}.$$

Figure 2.18b shows simulations of the expression levels over time for the modified circuits. We see now that the expression levels all reach the same steady state value. The negative autoregulated circuit has the property that it reaches the steady state more quickly, due to the increased rate of protein expression when A is small ($\alpha'_B > \alpha_B$). Conversely, the positive autoregulated circuit has a slower rate of expression than the constitutive case, since we have lowered the rate of protein expression when A is small. The initial higher and lower expression rates are compensated for via the autoregulation, resulting in the same expression level in steady state. ▽

We have described how a Hill function can model the regulation of a gene by a single transcription factor. However, genes can also be regulated by multiple transcription factors, some of which may be activators and some may be repressors, as in the case of combinatorial promoters. The mRNA production rate can thus take several forms depending on the roles (activators versus repressors) of the various transcription factors. In general, the production rate resulting from a promoter that takes as input transcription factors P_i for $i \in \{1, ..., N\}$ will be denoted $F(P_1, ..., P_N)$.

The dynamics of a transcriptional module are often well captured by the ordinary differential equations

$$\frac{dm_{P_i}}{dt} = F(P_1, ..., P_N) - \delta_{P_i} m_{P_i}, \qquad \frac{dP_i}{dt} = \kappa_{P_i} m_{P_i} - \gamma_{P_i} P_i. \qquad (2.24)$$

For a combinatorial promoter with two input proteins, an activator P_a and a repressor P_r, in which, for example, the activator cannot bind if the repressor is bound to the promoter, the function $F(P_a, P_r)$ can be obtained by employing the competitive binding in the reduced-order models of Section 2.1. In this case, assuming the activator has cooperativity n and the repressor has cooperativity m, we obtain the expression

$$F(P_a, P_r) = \alpha \frac{(P_a/K_a)^n}{1 + (P_a/K_a)^n + (P_r/K_r)^m}, \qquad (2.25)$$

where $K_a = (K_{m,a} K_{d,a})^{(1/n)}$, $K_r = (K_{m,r} K_{d,r})^{(1/m)}$, in which $K_{d,a}$ and $K_{d,r}$ are the dissociation constants of the activator and repressor, respectively, from the DNA promoter site, while $K_{m,a}$ and $K_{m,r}$ are the dissociation constants for the cooperative binding reactions for the activator and repressor, respectively. In these expressions, RNA polymerase does not explicitly appear as it affects the values of the dissociation constants and of α. In the case in which the activator is "leaky," that is,

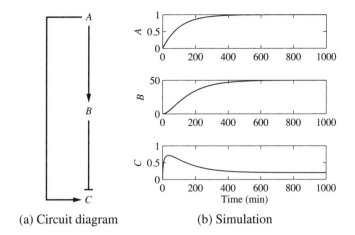

(a) Circuit diagram   (b) Simulation

Figure 2.19: The incoherent feedforward loop (type I). (a) A schematic diagram of the circuit. (b) A simulation of the model in equation (2.28) with $\beta_A = 0.01$ μM/min, $\gamma = 0.01$ min^{-1}, $\beta_B = 1$ μM/min, $\beta_C = 100$ μM/min, $K_B = 0.001$ μM , and $K_A = 1$ μM.

some transcription still occurs even when there is no activator, the above expression should be modified to

$$F(P_a, P_r) = \alpha \frac{(P_a/K_a)^n}{1 + (P_a/K_a)^n + (P_r/K_r)^m} + \alpha_0, \qquad (2.26)$$

where α_0 is the basal transcription rate when no activator is present. If the basal rate can still be repressed by the repressor, the above expression should be modified to (see Exercise 2.10)

$$F(P_a, P_r) = \frac{\alpha(P_a/K_a)^n + \alpha_0}{1 + (P_a/K_a)^n + (P_r/K_r)^m}. \qquad (2.27)$$

Example 2.3 (Incoherent feedforward loops). Combinatorial promoters with two inputs are often used in systems where a logical "and" is required. As an example, we illustrate here an incoherent feedforward loop (type I) [4]. Such a circuit is composed of three transcription factors A, B, and C, in which A directly activates C and B while B represses C. This is illustrated in Figure 2.19a. This is different from a coherent feedforward loop in which both A and B activate C. In the incoherent feedforward loop, if we would like C to be high only when A is high and B is low ("and" gate), we can consider a combinatorial promoter in which the activator A and the repressor B competitively bind to the promoter of C. The resulting Hill function is given by the expression in equation (2.25). Depending on the values of the constants, the expression of C is low unless A is high and B is low. The resulting ODE model, neglecting the mRNA dynamics, is given by the system

$$\frac{dA}{dt} = \beta_A - \gamma A,$$

$$\frac{dB}{dt} = \beta_B \frac{A/K_A}{1 + (A/K_A)} - \gamma B, \qquad (2.28)$$

$$\frac{dC}{dt} = \beta_C \frac{A/K_A}{1 + (A/K_A) + (B/K_B)} - \gamma C,$$

in which we have assumed no cooperativity of binding for both the activator and the repressor. If we view β_A as an input to the system and C as an output, we can investigate how this output responds to a sudden increase of β_A. Upon a sudden increase of β_A, protein A builds up and binds to the promoter of C initiating transcription, so that protein C starts getting produced. At the same time, protein B is produced and accumulates until it reaches a large enough value to repress C. Hence, we can expect a pulse of C production for suitable parameter values. This is shown in Figure 2.19b. Note that if the production rate constant β_C is very large, a little amount of A will cause C to immediately tend to a very high concentration. This explains the large initial slope of the C signal in Figure 2.19b.   ▽

2.4 Post-transcriptional regulation

In addition to regulation of expression through modifications of the process of transcription, cells can also regulate the production and activity of proteins via a collection of other post-transcriptional modifications. These include methods of modulating the translation of proteins, as well as affecting the activity of a protein via changes in its conformation.

Allosteric modifications to proteins

In allosteric regulation, a regulatory molecule, called allosteric effector, binds to a site separate from the catalytic site (active site) of an enzyme. This binding causes a change in the conformation of the protein, turning off (or turning on) the catalytic site (Figure 2.20).

An allosteric effector can either be an activator or an inhibitor, just like inducers work for activation or inhibition of transcription factors. Inhibition can either be competitive or not competitive. In the case of competitive inhibition, the inhibitor competes with the substrate for binding the enzyme, that is, the substrate can bind to the enzyme only if the inhibitor is not bound. In the case of non-competitive inhibition, the substrate can be bound to the enzyme even if the latter is bound to the inhibitor. In this case, however, the product may not be able to form or may form at a lower rate, in which case, we have partial inhibition.

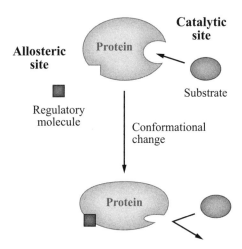

Figure 2.20: In allosteric regulation, a regulatory molecule binds to a site separate from the catalytic site (active site) of an enzyme. This binding causes a change in the three-dimensional conformation of the protein, turning off (or turning on) the catalytic site. Figure adapted from http://courses.washington.edu/conj/protein/proregulate.htm

Activation can be absolute or not. Specifically, an allosteric effector is an *absolute activator* when the enzyme can bind to the substrate only when the enzyme is bound to the allosteric effector. Otherwise, the allosteric effector is a non-absolute activator. In this section, we derive the expressions for the production rate of the active protein in an enzymatic reaction in the two most common cases: when we have a (non-competitive) inhibitor I or an (absolute) activator A of the enzyme.

Allosteric inhibition

Consider the standard enzymatic reaction

$$E + S \underset{d}{\overset{a}{\rightleftharpoons}} ES \overset{k}{\rightarrow} E + P,$$

in which enzyme E binds to substrate S and transforms it into the product P. Let I be a (non-competitive) inhibitor of enzyme E so that when E is bound to I, the complex EI can still bind to substrate S, however, the complex EIS is non-productive, that is, it does not produce P. Then, we have the following additional reactions:

$$E + I \underset{k_-}{\overset{k_+}{\rightleftharpoons}} EI, \qquad ES + I \underset{k_-}{\overset{k_+}{\rightleftharpoons}} EIS, \qquad EI + S \underset{d}{\overset{a}{\rightleftharpoons}} EIS,$$

in which, for simplicity of notation, we have assumed that the dissociation constant between E and I does not depend on whether E is bound to the substrate S. Similarly, we have assumed that the dissociation constant of E from S does not depend on whether the inhibitor I is bound to E. Additionally, we have the conservation laws:

$$E_{tot} = E + [ES] + [EI] + [EIS], \qquad S_{tot} = S + P + [ES] + [EIS].$$

The production rate of P is given by $dP/dt = k[ES]$. Since binding reactions are very fast, we can assume that all the complexes' concentrations are at the quasi-steady state. This gives

$$[EIS] = \frac{a}{d}[EI] \cdot S, \qquad [EI] = \frac{k_+}{k_-}E \cdot I, \qquad [ES] = \frac{S \cdot E}{K_m},$$

where $K_m = (d + k)/a$ is the Michaelis-Menten constant. Using these expressions, the conservation law for the enzyme, and the fact that $a/d \approx 1/K_m$, we obtain

$$E = \frac{E_{tot}}{(I/K_d + 1)(1 + S/K_m)}, \quad \text{with } K_d = k_-/k_+,$$

so that

$$[ES] = \frac{S}{S + K_m}\frac{E_{tot}}{1 + I/K_d}$$

and, as a consequence,

$$\frac{dP}{dt} = kE_{tot}\left(\frac{1}{1 + I/K_d}\right)\left(\frac{S}{S + K_m}\right).$$

In our earlier derivations of the Michaelis-Menten kinetics $V_{max} = kE_{tot}$ was called the maximal velocity, which occurs when the enzyme is completely saturated by the substrate (Section 2.1, equation (2.9)). Hence, the effect of a non-competitive inhibitor is to decrease the maximal velocity V_{max} to $V_{max}/(1 + I/K_d)$.

Another type of inhibition occurs when the inhibitor is competitive, that is, when I is bound to E, the complex EI cannot bind to protein S. Since E can either bind to I or S (not both), I competes against S for binding to E (see Exercise 2.13).

Allosteric activation

In this case, the enzyme E can transform S to its active form only when it is bound to A. Also, we assume that E cannot bind S unless E is bound to A (from here, the name absolute activator). The reactions should be modified to

$$E + A \underset{k_-}{\overset{k_+}{\rightleftharpoons}} EA, \qquad EA + S \underset{d}{\overset{a}{\rightleftharpoons}} EAS \overset{k}{\to} P + EA,$$

with conservation laws

$$E_{tot} = E + [EA] + [EAS], \qquad S_{tot} = S + P + [EAS].$$

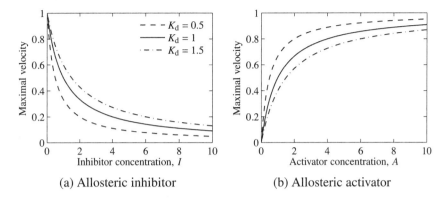

(a) Allosteric inhibitor   (b) Allosteric activator

Figure 2.21: Maximal velocity in the presence of allosteric effectors (inhibitors or activators). The plots in (a) show the maximal velocity $V_{max}/(1 + I/K_d)$ as a function of the inhibitor concentration I. The plots in (b) show the maximal velocity $V_{max}A/(A + K_d)$ as a function of the activator concentration A. The different plots show the effect of the dissociation constant for $V_{max} = 1$.

The production rate of P is given by $dP/dt = k\,[EAS]$. Assuming as above that the complexes are at the quasi-steady state, we have that

$$[EA] = \frac{E \cdot A}{K_d}, \qquad [EAS] = \frac{S \cdot [EA]}{K_m},$$

which, using the conservation law for E, leads to

$$E = \frac{E_{tot}}{(1 + S/K_m)(1 + A/K_d)} \quad \text{and} \quad [EAS] = \left(\frac{A}{A + K_d}\right)\left(\frac{S}{S + K_m}\right)E_{tot}.$$

Hence, we have that

$$\frac{dP}{dt} = kE_{tot}\left(\frac{A}{A + K_d}\right)\left(\frac{S}{S + K_m}\right).$$

The effect of an absolute activator is to modulate the maximal speed of modification by a factor $A/(A + K_d)$.

Figure 2.21 shows the behavior of the maximal velocity as a function of the allosteric effector concentration. As the dissociation constant decreases, that is, the affinity of the effector increases, a very small amount of effector will cause the maximal velocity to reach V_{max} in the case of the activator and 0 in the case of the inhibitor.

Another type of activation occurs when the activator is not absolute, that is, when E can bind to S directly, but cannot activate S unless the complex ES first binds A (see Exercise 2.14).

Covalent modifications to proteins

In addition to regulation that controls transcription of DNA into mRNA, a variety of mechanisms are available for controlling expression after mRNA is produced. These include control of splicing and transport from the nucleus (in eukaryotes), the use of various secondary structure patterns in mRNA that can interfere with ribosomal binding or cleave the mRNA into multiple pieces, and targeted degradation of mRNA. Once the polypeptide chain is formed, additional mechanisms are available that regulate the folding of the protein as well as its shape and activity level.

One of the most common types of post-transcriptional regulation is through the covalent modification of proteins, such as through the process of *phosphorylation*. Phosphorylation is an enzymatic process in which a phosphate group is added to a protein and the resulting conformation of the protein changes, usually from an inactive configuration to an active one. The enzyme that adds the phosphate group is called a *kinase* and it operates by transferring a phosphate group from a bound ATP molecule to the protein, leaving behind ADP and the phosphorylated protein. *Dephosphorylation* is a complementary enzymatic process that can remove a phosphate group from a protein. The enzyme that performs dephosphorylation is called a *phosphatase*. Figure 2.3 shows the process of phosphorylation in more detail.

Since phosphorylation and dephosphorylation can occur much more quickly than protein production and degradation, it is used in biological circuits in which a rapid response is required. One common pattern is that a signaling protein will bind to a ligand and the resulting allosteric change allows the signaling protein to serve as a kinase. The newly active kinase then phosphorylates a second protein, which modulates other functions in the cell. Phosphorylation cascades can also be used to amplify the effect of the original signal; we will describe this in more detail in Section 2.5.

Kinases in cells are usually very specific to a given protein, allowing detailed signaling networks to be constructed. Phosphatases, on the other hand, are much less specific, and a given phosphatase species may dephosphorylate many different types of proteins. The combined action of kinases and phosphatases is important in signaling since the only way to deactivate a phosphorylated protein is by removing the phosphate group. Thus phosphatases are constantly "turning off" proteins, and the protein is activated only when sufficient kinase activity is present.

Phosphorylation of a protein occurs by the addition of a charged phosphate (PO_4) group to the serine (Ser), threonine (Thr) or tyrosine (Tyr) amino acids. Similar covalent modifications can occur by the attachment of other chemical groups to select amino acids. *Methylation* occurs when a methyl group (CH_3) is added to lysine (Lys) and is used for modulation of receptor activity and in modifying histones that are used in chromatin structures. *Acetylation* occurs when an acetyl group ($COCH_3$) is added to lysine and is also used to modify histones. *Ubiquitina-*

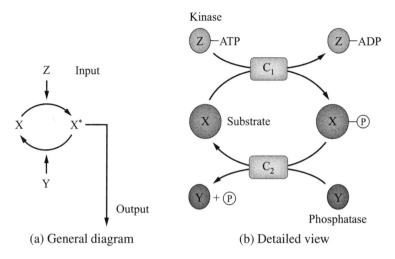

Figure 2.22: (a) General diagram representing a covalent modification cycle. (b) Detailed view of a phosphorylation cycle including ATP, ADP, and the exchange of the phosphate group "p."

tion refers to the addition of a small protein, ubiquitin, to lysine; the addition of a polyubiquitin chain to a protein targets it for degradation.

Here, we focus on *reversible* cycles of modification, in which a protein is interconverted between two forms that differ in activity. At a high level, a covalent modification cycle involves a target protein X, an enzyme Z for modifying it, and a second enzyme Y for reversing the modification (see Figure 2.22). We call X* the activated protein. There are often allosteric effectors or further covalent modification systems that regulate the activity of the modifying enzymes, but we do not consider this added level of complexity here. The reactions describing this system are given by the following two enzymatic reactions, also called a *two-step reaction model*:

$$Z+X \underset{d_1}{\overset{a_1}{\rightleftharpoons}} C_1 \overset{k_1}{\rightarrow} X^* + Z, \qquad Y+X^* \underset{d_2}{\overset{a_2}{\rightleftharpoons}} C_2 \overset{k_2}{\rightarrow} X+Y,$$

in which we have let C_1 be the kinase/protein complex and C_2 be the active protein/phosphatase complex. The corresponding differential equation model is given by

$$\frac{dZ}{dt} = -a_1 Z \cdot X + (k_1 + d_1)C_1, \qquad \frac{dX^*}{dt} = k_1 C_1 - a_2 Y \cdot X^* + d_2 C_2,$$

$$\frac{dX}{dt} = -a_1 Z \cdot X + d_1 C_1 + k_2 C_2, \qquad \frac{dC_2}{dt} = a_2 Y \cdot X^* - (d_2 + k_2)C_2,$$

$$\frac{dC_1}{dt} = a_1 Z \cdot X - (d_1 + k_1)C_1, \qquad \frac{dY}{dt} = -a_2 Y \cdot X^* + (d_2 + k_2)C_2.$$

Furthermore, we have that the total amounts of enzymes Z and Y are conserved. Denote the total concentrations of Z, Y, and X by Z_{tot}, Y_{tot}, and X_{tot}, respectively.

Then, we have also the conservation laws

$$Z + C_1 = Z_{\text{tot}}, \qquad Y + C_2 = Y_{\text{tot}}, \qquad X + X^* + C_1 + C_2 = X_{\text{tot}}.$$

Using the first two conservation laws, we can reduce the above system of differential equations to the following one:

$$\frac{dC_1}{dt} = a_1(Z_{\text{tot}} - C_1) \cdot X - (d_1 + k_1)C_1,$$

$$\frac{dX^*}{dt} = k_1 C_1 - a_2(Y_{\text{tot}} - C_2) \cdot X^* + d_2 C_2,$$

$$\frac{dC_2}{dt} = a_2(Y_{\text{tot}} - C_2) \cdot X^* - (d_2 + k_2)C_2.$$

As in the case of the enzymatic reaction, this system cannot be analytically integrated. To simplify it, we can perform a similar approximation as done for the enzymatic reaction. In particular, the complexes' concentrations C_1 and C_2 reach their steady state values very quickly under the assumption $a_1 Z_{\text{tot}}, a_2 Y_{\text{tot}}, d_1, d_2 \gg k_1, k_2$. Therefore, we can approximate the above system by substituting for C_1 and C_2 their steady state values, given by the solutions to

$$a_1(Z_{\text{tot}} - C_1) \cdot X - (d_1 + k_1)C_1 = 0$$

and

$$a_2(Y_{\text{tot}} - C_2) \cdot X^* - (d_2 + k_2)C_2 = 0.$$

By solving these equations, we obtain that

$$C_2 = \frac{Y_{\text{tot}} X^*}{X^* + K_{m,2}}, \quad \text{with} \quad K_{m,2} = \frac{d_2 + k_2}{a_2},$$

and

$$C_1 = \frac{Z_{\text{tot}} X}{X + K_{m,1}}, \quad \text{with} \quad K_{m,1} = \frac{d_1 + k_1}{a_1}.$$

As a consequence, the model of the phosphorylation system can be well approximated by

$$\frac{dX^*}{dt} = k_1 \frac{Z_{\text{tot}} X}{X + K_{m,1}} - a_2 \frac{Y_{\text{tot}} K_{m,2}}{X^* + K_{m,2}} \cdot X^* + d_2 \frac{Y_{\text{tot}} X^*}{X^* + K_{m,2}},$$

which, considering that $a_2 K_{m,2} - d_2 = k_2$, leads finally to

$$\frac{dX^*}{dt} = k_1 \frac{Z_{\text{tot}} X}{X + K_{m,1}} - k_2 \frac{Y_{\text{tot}} X^*}{X^* + K_{m,2}}. \tag{2.29}$$

We will come back to the modeling of this system after we have introduced singular perturbation theory, through which we will be able to perform a formal analysis and mathematically characterize the assumptions needed for approximating the original system by the first-order model (2.29). Also, note that X should be replaced by using the conservation law by $X = X_{tot} - X^* - C_1 - C_2$, which can be solved for X using the expressions of C_1 and C_2. Under the common assumption that the amount of enzyme is much smaller than the amount of substrate ($Z_{tot}, Y_{tot} \ll X_{tot}$) [36], we have that $X \approx X_{tot} - X^*$ [36], leading to a form of the differential equation (2.29) that is simple enough to be analyzed mathematically.

Simpler models of phosphorylation cycles can be considered, which oftentimes are instructive as a first step to study a specific question of interest. In particular, the one-step reaction model neglects the complex formation in the two enzymatic reactions and simply models them as a single irreversible reaction (see Exercise 2.12).

It is important to note that the speed of enzymatic reactions, such as phosphorylation and dephosphorylation, is usually much faster than the speed of protein production and protein decay. In particular, the values of the catalytic rate constants k_1 and k_2, even if changing greatly from organism to organism, are typically several orders of magnitude larger than protein decay and can be on the order of 10^3 min^{-1} in bacteria where typical rates of protein decay are about 0.01 min^{-1} (http://bionumbers.hms.harvard.edu/).

Ultrasensitivity

One relevant aspect of the response of the covalent modification cycle to its input is the sensitivity of the steady state characteristic curve, that is, the map that determines the equilibrium value of the output X^* corresponding to a value of the input Z_{tot}. Specifically, which parameters affect the shape of the steady state characteristic is a crucial question. To study this, we set $dX^*/dt = 0$ in equation (2.29). Using the approximation $X \approx X_{tot} - X^*$, defining $\bar{K}_1 := K_{m,1}/X_{tot}$ and $\bar{K}_2 := K_{m,2}/X_{tot}$, we obtain

$$y := \frac{k_1 Z_{tot}}{k_2 Y_{tot}} = \frac{X^*/X_{tot}\left(\bar{K}_1 + (1 - X^*/X_{tot})\right)}{(\bar{K}_2 + X^*/X_{tot})(1 - X^*/X_{tot})}. \tag{2.30}$$

Since y is proportional to the input Z_{tot}, we study the equilibrium value of X^* as a function of y. This function is usually characterized by two key parameters: the response coefficient, denoted R, and the point of half maximal induction, denoted y_{50}. Let y_α denote the value of y corresponding to having X^* equal $\alpha\%$ of the maximum value of X^* obtained for $y = \infty$, which is equal to X_{tot}. Then, the response coefficient is defined as

$$R := \frac{y_{90}}{y_{10}},$$

and measures how switch-like the response of X^* is to changes in y (Figure 2.23). When $R \to 1$ the response becomes switch-like. In the case in which the steady state characteristic is a Hill function, we have that $X^* = (y/K)^n/(1 + (y/K)^n)$, so

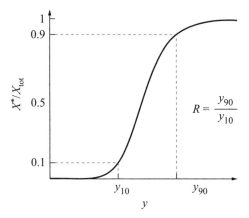

Figure 2.23: Steady state characteristic curve showing the relevance of the response coefficient for ultrasensitivity. As $R \to 1$, the points y_{10} and y_{90} tend to each other.

that $y_\alpha = (\alpha/(100-\alpha))^{(1/n)}$ and as a consequence

$$R = (81)^{(1/n)}, \quad \text{or equivalently} \quad n = \frac{\log(81)}{\log(R)}.$$

Hence, when $n = 1$, that is, the characteristic is of the Michaelis-Menten type, we have that $R = 81$, while when n increases, R decreases. Usually, when $n > 1$ the response is referred to as *ultrasensitive* and the formula $n = \log(81)/\log(R)$ is often employed to estimate the *apparent Hill coefficient* of an experimentally obtained steady state characteristic since R can be calculated directly from the data points.

In the case of the current system, from equation (2.30), we have that

$$y_{90} = \frac{(\bar{K}_1 + 0.1)\,0.9}{(\bar{K}_2 + 0.9)\,0.1} \quad \text{and} \quad y_{10} = \frac{(\bar{K}_1 + 0.9)\,0.1}{(\bar{K}_2 + 0.1)\,0.9},$$

so that

$$R = 81 \frac{(\bar{K}_1 + 0.1)(\bar{K}_2 + 0.1)}{(\bar{K}_2 + 0.9)(\bar{K}_1 + 0.9)}. \tag{2.31}$$

As a consequence, when $\bar{K}_1, \bar{K}_2 \gg 1$, we have that $R \to 81$, which gives a Michaelis-Menten type of response. If instead $\bar{K}_1, \bar{K}_2 \ll 0.1$, we have that $R \to 1$, which corresponds to a theoretical Hill coefficient $n \gg 1$, that is, a switch-like response (Figure 2.24). In particular, if we have, for example, $\bar{K}_1 = \bar{K}_2 = 10^{-2}$, we obtain an apparent Hill coefficient greater than 13. This type of ultrasensitivity is usually referred to as *zero-order ultrasensitivity*. The reason for this name is due to the fact that when $K_{m,1}$ is much smaller than the total amount of protein substrate X_{tot}, we have that $Z_{tot}X/(K_{m,1} + X) \approx Z_{tot}$. Hence, the forward modification rate is "zero order" in the substrate concentration (no free enzyme is left, all is bound to the substrate).

One can study the behavior also of the point of half maximal induction

$$y_{50} = \frac{\bar{K}_1 + 0.5}{\bar{K}_2 + 0.5},$$

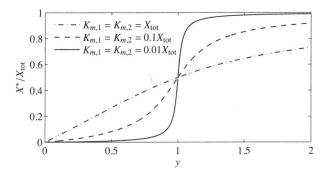

Figure 2.24: Steady state characteristic curve of a covalent modification cycle as a function of the Michaelis-Menten constants $K_{m,1}$ and $K_{m,2}$.

to find that as $\bar{K}_2$ increases, y_{50} decreases and that as $\bar{K}_1$ increases, y_{50} increases.

Phosphotransfer systems

Phosphotransfer systems are also a common motif in cellular signal transduction. These structures are composed of proteins that can phosphorylate each other. In contrast to kinase-mediated phosphorylation, where the phosphate donor is usually ATP, in phosphotransfer the phosphate group comes from the donor protein itself (Figure 2.25). Each protein carrying a phosphate group can donate it to the next protein in the system through a reversible reaction.

Let X be a protein in its inactive form and let X^* be the same protein once it has been activated by the addition of a phosphate group. Let Z^* be a phosphate donor, that is, a protein that can transfer its phosphate group to the acceptor X. The standard phosphotransfer reactions can be modeled according to the two-step reaction model

$$Z^* + X \underset{k_2}{\overset{k_1}{\rightleftharpoons}} C_1 \underset{k_4}{\overset{k_3}{\rightleftharpoons}} X^* + Z,$$

in which C_1 is the complex of Z bound to X bound to the phosphate group. Additionally, we assume that protein Z can be phosphorylated and protein X^* dephosphorylated by other phosphorylation reactions by which the phosphate group is taken to and removed from the system. These reactions are modeled as one-step reactions depending only on the concentrations of Z and X^*, that is:

$$Z \xrightarrow{\pi_1} Z^*, \qquad X^* \xrightarrow{\pi_2} X.$$

Proteins X and Z are conserved in the system, that is, $X_{\text{tot}} = X + C_1 + X^*$ and $Z_{\text{tot}} = Z + C_1 + Z^*$. We view the total amount of Z, Z_{tot}, as the input to our system and the amount of phosphorylated form of X, X^*, as the output. We are interested in the steady state characteristic curve describing how the steady state value of X^* depends on the value of Z_{tot}.

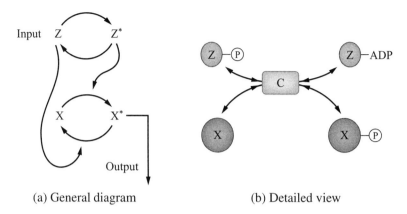

(a) General diagram   (b) Detailed view

Figure 2.25: (a) Diagram of a phosphotransfer system. (b) Proteins X and Z are transferring the phosphate group p to each other.

The differential equation model corresponding to this system is given by the equations

$$\frac{dC_1}{dt} = k_1 \left(X_{\text{tot}} - X^* - C_1\right) \cdot Z^* - k_3 C_1 - k_2 C_1 + k_4 X^* \cdot (Z_{\text{tot}} - C_1 - Z^*),$$

$$\frac{dZ^*}{dt} = \pi_1(Z_{\text{tot}} - C_1 - Z^*) + k_2 C_1 - k_1 \left(X_{\text{tot}} - X^* - C_1\right) \cdot Z^*, \qquad (2.32)$$

$$\frac{dX^*}{dt} = k_3 C_1 - k_4 X^* \cdot (Z_{\text{tot}} - C_1 - Z^*) - \pi_2 X^*.$$

The steady state transfer curve is shown in Figure 2.26 and it is obtained by simulating system (2.32) and recording the equilibrium values of X^* corresponding to different values of Z_{tot}. The transfer curve is linear for a large range of values of Z_{tot} and can be rendered fairly close to a linear relationship for values of Z_{tot}

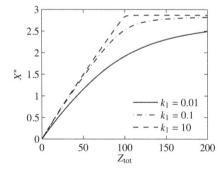

Figure 2.26: Steady state characteristic curve of the phosphotransfer system. Here, we have set $k_2 = k_3 = 0.1$ s^{-1}, $k_4 = 0.1$ nM^{-1} s^{-1}, $\pi_1 = \pi_2 = 3.1$ s^{-1}, and $X_{\text{tot}} = 100$ nM.

smaller than X_{tot} by increasing k_1. The slope of this linear relationship can be further tuned by changing the values of k_3 and k_4 (see Exercise 2.15). Hence, this system can function as an approximately linear anplifier. Its use in the realization of insulation devices that attenuate the effects of loading from interconnections will be illustrated in Chapter 6.

2.5 Cellular subsystems

In the previous section we have studied how to model a variety of core processes that occur in cells. In this section we consider a few common "subsystems" in which these processes are combined for specific purposes.

Intercellular signaling: MAPK cascades

The mitogen activated protein kinase (MAPK) cascade is a recurrent structural motif in several signal transduction pathways (Figure 2.27). The cascade consists of a MAPK kinase (MAPKKK), denoted X_0, a MAPK kinase (MAPKK), denoted X_1, and a MAPK, denoted X_2. MAPKKKs activate MAPKKs by phosphorylation at two conserved sites and MAPKKs activate MAPKs by phosphorylation at conserved sites. The cascade relays signals from the cell membrane to targets in the cytoplasm and nucleus. It has been extensively studied and modeled. Here, we provide a model for double phosphorylation, which is one of the main building blocks of the MAPK cascade. Then, we construct a detailed model of the MAPK cascade, including the reactions describing each stage and the corresponding rate equations.

Double phosphorylation model. Consider the double phosphorylation motif in Figure 2.28. The reactions describing the system are given by

$$E_1 + X \underset{d_1}{\overset{a_1}{\rightleftharpoons}} C_1 \xrightarrow{k_1} X^* + E_1, \qquad E_2 + X^* \underset{d_2}{\overset{a_2}{\rightleftharpoons}} C_2 \xrightarrow{k_2} X + E_2,$$

$$X^* + E_1 \underset{d_1^*}{\overset{a_1^*}{\rightleftharpoons}} C_3 \xrightarrow{k_1^*} X^{**} + E_1, \qquad E_2 + X^{**} \underset{d_2^*}{\overset{a_2^*}{\rightleftharpoons}} C_4 \xrightarrow{k_2^*} X^* + E_2,$$

in which C_1 is the complex of E_1 with X, C_2 is the complex of E_2 with X^*, C_3 is the complex of E_1 with X^*, and C_4 is the complex of E_2 with X^{**}. The conservation laws are given by

$$E_1 + C_1 + C_3 = E_{1,tot}, \qquad E_2 + C_2 + C_4 = E_{2,tot},$$
$$X_{tot} = X + X^* + X^{**} + C_1 + C_2 + C_3 + C_4.$$

As performed earlier, we assume that the complexes are at the quasi-steady state since binding reactions are very fast compared to the catalytic reactions. This gives

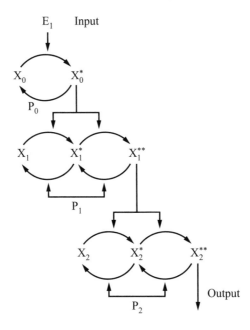

Figure 2.27: Schematic representation of the MAPK cascade. It has three levels: the first one has a single phosphorylation, while the second and the third ones have a double phosphorylation.

the Michaelis-Menten form for the amount of formed complexes:

$$C_1 = E_{1,\text{tot}} \frac{K_1^* X}{K_1^* X + K_1 X^* + K_1 K_1^*}, \qquad C_3 = E_{1,\text{tot}} \frac{K_1 X^*}{K_1^* X + K_1 X^* + K_1 K_1^*},$$

$$C_2 = E_{2,\text{tot}} \frac{K_2^* X^*}{K_2^* X^* + K_2 X^{**} + K_2 K_2^*}, \qquad C_4 = E_{2,\text{tot}} \frac{K_2 X^{**}}{K_2^* X^* + K_2 X^{**} + K_2 K_2^*},$$

in which $K_i = (d_i + k_i)/a_i$ and $K_i^* = (d_i^* + k_i^*)/a_i^*$ are the Michaelis-Menten constants for the enzymatic reactions. Since the complexes are at the quasi-steady state, it

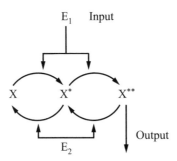

Figure 2.28: Schematic representation of a double phosphorylation cycle. E_1 is the input and X^{**} is the output.

follows that

$$\frac{d}{dt}X^* = k_1 C_1 - k_2 C_2 - k_1^* C_3 + k_2^* C_4,$$

$$\frac{d}{dt}X^{**} = k_1^* C_3 - k_2^* C_4,$$

from which, substituting the expressions of the complexes, we obtain that

$$\frac{d}{dt}X^* = E_{1,\text{tot}}\frac{k_1 X K_1^* - k_1^* X^* K_1}{K_1^* X + K_1 X^* + K_1^* K_1} + E_{2,\text{tot}}\frac{k_2^* X^{**} K_2 - k_2 X^* K_2^*}{K_2^* X^* + K_2 X^{**} + K_2 K_2^*},$$

$$\frac{d}{dt}X^{**} = k_1^* E_{1,\text{tot}}\frac{K_1 X^*}{K_1^* X + K_1 X^* + K_1 K_1^*} - k_2^* E_{2,\text{tot}}\frac{K_2 X^{**}}{K_2^* X^* + K_2 X^{**} + K_2 K_2^*}.$$

Detailed model of the MAPK cascade. We now give the entire set of reactions for the MAPK cascade of Figure 2.27 as they are found in standard references (Huang-Ferrell model [44]):

$$E_1 + X_0 \xrightleftharpoons[d_{1,0}]{a_{1,0}} C_1 \xrightarrow{k_{1,0}} X_0^* + E_1, \qquad P_0 + X_0^* \xrightleftharpoons[d_{2,0}]{a_{2,0}} C_2 \xrightarrow{k_{2,0}} X_0 + P_0,$$

$$X_0^* + X_1 \xrightleftharpoons[d_{1,1}]{a_{1,1}} C_3 \xrightarrow{k_{1,1}} X_1^* + X_0^*, \qquad X_1^* + P_1 \xrightleftharpoons[d_{2,1}]{a_{2,1}} C_4 \xrightarrow{k_{2,1}} X_1 + P_1,$$

$$X_0^* + X_1^* \xrightleftharpoons[d_{1,1}^*]{a_{1,1}^*} C_5 \xrightarrow{k_{1,1}^*} X_1^{**} + X_0^*, \qquad X_1^{**} + P_1 \xrightleftharpoons[d_{2,1}^*]{a_{2,1}^*} C_6 \xrightarrow{k_{2,1}^*} X_1^* + P_1,$$

$$X_1^{**} + X_2 \xrightleftharpoons[d_{1,2}]{a_{1,2}} C_7 \xrightarrow{k_{1,2}} X_2^* + X_1^{**}, \qquad X_2^* + P_2 \xrightleftharpoons[d_{2,2}]{a_{2,2}} C_8 \xrightarrow{k_{2,2}} X_2 + P_2,$$

$$X_1^{**} + X_2^* \xrightleftharpoons[d_{1,2}^*]{a_{1,2}^*} C_9 \xrightarrow{k_{1,2}^*} X_2^{**} + X_1^{**}, \qquad X_2^{**} + P_2 \xrightleftharpoons[d_{2,2}^*]{a_{2,2}^*} C_{10} \xrightarrow{k_{2,2}^*} X_2^* + P_2,$$

with conservation laws

$$X_{0,\text{tot}} = X_0 + X_0^* + C_1 + C_2 + C_3 + C_5,$$

$$X_{1,\text{tot}} = X_1 + X_1^* + C_3 + X_1^{**} + C_4 + C_5 + C_6 + C_7 + C_9,$$

$$X_{2,\text{tot}} = X_2 + X_2^* + X_2^{**} + C_7 + C_8 + C_9 + C_{10},$$

$$E_{1,\text{tot}} = E_1 + C_1, \quad P_{0,\text{tot}} = P_0 + C_2,$$

$$P_{1,\text{tot}} = P_1 + C_4 + C_6,$$

$$P_{2,\text{tot}} = P_2 + C_8 + C_{10}.$$

The corresponding ODE model is given by

$$\frac{dC_1}{dt} = a_{1,0} E_1 X_0 - (d_{1,0} + k_{1,0})\, C_1,$$

$$\frac{dX_0^*}{dt} = k_{1,0}\, C_1 + d_{2,0}\, C_2 - a_{2,0}\, P_0\, X_0^* + (d_{1,1} + k_{1,1})\, C_3 - a_{1,1}\, X_1\, X_0^*$$
$$+ (d_{1,1}^* + k_{1,1}^*)\, C_5 - a_{1,1}^*\, X_0^*\, X_1^*,$$

$$\frac{dC_2}{dt} = a_{2,0}\, P_0\, X_0^* - (d_{2,0} + k_{2,0})\, C_2,$$

$$\frac{dC_3}{dt} = a_{1,1}\, X_1\, X_0^* - (d_{1,1} + k_{1,1})\, C_3,$$

$$\frac{dX_1^*}{dt} = k_{1,1}\, C_3 + d_{2,1}\, C_4 - a_{2,1}\, X_1^*\, P_1 + d_{1,1}^* C_5 - a_{1,1}^*\, X_1^*\, X_0^* + k_{2,1}^*\, C_6,$$

$$\frac{dC_4}{dt} = a_{2,1}\, X_1^*\, P_1 - (d_{2,1} + k_{2,1})\, C_4,$$

$$\frac{dC_5}{dt} = a_{1,1}^*\, X_0^*\, X_1^* - (d_{1,1}^* + k_{1,1}^*)\, C_5,$$

$$\frac{dX_1^{**}}{dt} = k_{1,1}^*\, C_5 - a_{2,1}^*\, X_1^*\, P_1 + d_{2,1}^*\, C_6 - a_{1,2}\, X_1^{**}\, X_2$$
$$+ (d_{1,2} + k_{1,2})\, C_7 - a_{1,2}^*\, X_1^{**}\, X_2^* + (d_{1,2}^* + k_{1,2}^*)\, C_9,$$

$$\frac{dC_6}{dt} = a_{2,1}^*\, X_1^{**}\, P_1 - (d_{2,1}^* + k_{2,1}^*)\, C_6,$$

$$\frac{dC_7}{dt} = a_{1,2}^*\, X_1^*\, X_2 - (d_{1,2}^* + k_{1,2}^*)\, C_7,$$

$$\frac{dX_2^*}{dt} = -a_{2,2}\, X_2^*\, P_2 + d_{2,2}\, C_8 - a_{1,2}^*\, X_2^*\, X_2^{**} + d_{1,2}^*\, C_9 + k_{2,2}^*\, C_{10},$$

$$\frac{dC_8}{dt} = a_{2,2}^*\, X_2^*\, P_2 - (d_{2,2} + k_{2,2})\, C_8,$$

$$\frac{dX_2^{**}}{dt} = k_{1,2}^*\, C_9 - a_{2,2}^*\, X_2^{**}\, P_2 + d_{2,2}^*\, C_{10},$$

$$\frac{dC_9}{dt} = a_{1,2}^*\, X_1^{**}\, X_2^* - (d_{1,2}^* + k_{1,2}^*)\, C_9,$$

$$\frac{dC_{10}}{dt} = a_{2,2}^*\, X_2^{**}\, P_2 - (d_{2,2}^* + k_{2,2}^*)\, C_{10}.$$

The steady state characteristic curve obtained with the mechanistic model predicts that the response of the MAPKKK to the stimulus $E_{1,\text{tot}}$ is of the Michaelis-Menten type. By contrast, the stimulus-response curve obtained for the MAPKK and MAPK are sigmoidal and show high Hill coefficients, which increase from the MAPKK response to the MAPK response. That is, an increased ultrasensitivity is observed moving down in the cascade (Figure 2.29). These model observations persist when key parameters, such as the Michaelis-Menten constants are changed [44]. Furthermore, zero-order ultrasensitivity effects can be observed. Specifically,

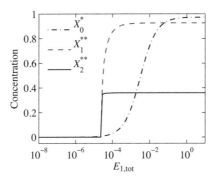

Figure 2.29: Steady state characteristic curve of the MAPK cascade for every stage. The x-axis shows concentration of $E_{1,\text{tot}}$ between 10^{-15} and 10^5 as indicated. Simulations from the model of [82].

if the amounts of MAPKK were increased, one would observe a higher apparent Hill coefficient for the response of MAPK. Similarly, if the values of the K_m for the reactions in which the MAPKK takes place were decreased, one would also observe a higher apparent Hill coefficient for the response of MAPK. Double phosphorylation is also key to obtain a high apparent Hill coefficient. In fact, a cascade in which the double phosphorylation was assumed to occur through a one-step model (similar to single phosphorylation) predicted substantially lower apparent Hill coefficients.

Notice that while phosphorylation cascades, such as the MAPK cascade, are usually viewed as unidirectional signal transmission systems, they actually allow information to travel backward (from downstream to upstream). This can be qualitatively seen as follows. Assuming as before that the total amounts of enzymes are much smaller than the total amounts of substrates ($E_{1,\text{tot}}, P_{0,\text{tot}}, P_{1,\text{tot}}, P_{2,\text{tot}} \ll X_{0,\text{tot}}, X_{1,\text{tot}}, X_{2,\text{tot}}$), we can approximate the conservation laws as

$$X_{0,\text{tot}} \approx X_0 + X_0^* + C_3 + C_5,$$
$$X_{1,\text{tot}} \approx X_1 + X_1^* + C_3 + X_1^{**} + C_5 + C_7 + C_9,$$
$$X_{2,\text{tot}} \approx X_2 + X_2^* + X_2^{**} + C_7 + C_9.$$

Using these and assuming that the complexes are at the quasi-steady state, we obtain the following functional dependencies:

$$C_1 = f_1(X_0^*, X_1^*, X_1^{**}, X_2^*, X_2^{**}), \qquad C_2 = f_2(X_0^*),$$
$$C_3 = f_3(X_0^*, X_1^*, X_1^{**}, X_2^*, X_2^{**}), \qquad C_5 = f_5(X_0^*, X_1^*),$$
$$C_7 = f_7(X_1^*, X_1^{**}, X_2^*, X_2^{**}), \qquad C_9 = f_9(X_1^{**}, X_2^*).$$

The fact that C_7 depends on X_2^* and X_2^{**} illustrates the counterintuitive fact that the dynamics of the second stage are influenced by those of the third stage. Similarly, the fact that C_3 depends on $X_1^*, X_1^{**}, X_2^*, X_2^{**}$ indicates that the dynamics of the first stage are influenced by those of the second stage and by that of the third stage. The phenomenon by which the behavior of a "module" is influenced by that of its downstream clients is called *retroactivity*, which is a phenomenon similar to loading in electrical and mechanical systems, studied at length in Chapter 6. This phenomenon in signaling cascades can allow perturbations to travel from downstream to upstream [77] and can lead to interesting dynamic behaviors such as sustained oscillations in the MAPK cascade [82].

Exercises

2.1 Consider a cascade of three activators $X \rightarrow Y \rightarrow Z$. Protein X is initially present in the cell in its inactive form. The input signal of X, S_x, appears at time $t = 0$. As a result, X rapidly becomes active and binds the promoter of gene Y, so that protein Y starts to be produced at rate β. When Y levels exceed a threshold K, gene Z begins to be transcribed and translated at rate β. All proteins have the same degradation/dilution rate γ.

   (i) What are the concentrations of proteins Y and Z as a function of time?

   (ii) What is the minimum duration of the pulse S_x such that Z will be produced?

   (iii) What is the response time of protein Z with respect to the time of addition of S_x?

2.2 (Switch-like behavior in cooperative binding) Derive the expressions of C and A as a function of B at the steady state when you have the cooperative binding reactions

$$B + B + \ldots + B \underset{k_2}{\overset{k_1}{\rightleftharpoons}} B_n, \quad B_n + A \underset{d}{\overset{a}{\rightleftharpoons}} C, \quad \text{and} \quad A + C = A_{\text{tot}}.$$

Make MATLAB plots of the expressions that you obtain and verify that as n increases the functions become more switch-like.

2.3 Consider the case of a competitive binding of an activator A and a repressor R with D and assume that before they can bind to D they have to cooperatively bind according to the following reactions:

$$A + A + \ldots + A \underset{k_2}{\overset{k_1}{\rightleftharpoons}} A_n, \quad R + R + \ldots + R \underset{\bar{k}_2}{\overset{\bar{k}_1}{\rightleftharpoons}} R_m,$$

in which the complex A_n contains n molecules of A and the complex R_m contains m molecules of R. The competitive binding reactions with A are given by

$$A_n + D \underset{d}{\overset{a}{\rightleftharpoons}} C, \quad R_m + D \underset{d'}{\overset{a'}{\rightleftharpoons}} C',$$

and $D_{tot} = D + C + C'$. What are the steady state expressions for C and D as functions of A and R?

2.4 Consider the following modification of the competitive binding reactions:

$$B_a + A \underset{d}{\overset{a}{\rightleftharpoons}} C, \qquad B_r + A \underset{\bar{d}}{\overset{\bar{a}}{\rightleftharpoons}} \bar{C}, \qquad C + B_r \underset{d'}{\overset{a'}{\rightleftharpoons}} C',$$

with $A_{tot} = A + C + \bar{C} + C'$. What are the steady state expressions for A and C? What information do you deduce from these expressions if A is a promoter, B_a is an activator protein, and C is the activator/DNA complex that makes the gene transcriptionally active?

2.5 Assume that we have an activator B_a and a repressor protein B_r. We want to obtain an input function such that when a large quantity of B_a is present, the gene is transcriptionally active only if there is no B_r, and when low amounts of B_a are present, the gene is transcriptionally inactive (with or without B_r). Write down the reactions among B_a, B_r, and the complexes formed with DNA (D) that lead to such an input function. Demonstrate that indeed the set of reactions you picked leads to the desired input function.

2.6 Consider the transcription and translation reactions incorporating the elongation process as considered in this chapter in equations (2.10)–(2.11). Modify them to the case in which an mRNA molecule can be translated to a polypeptide chain even while it is still elongating.

2.7 (Transcriptional regulation with delay) Consider a repressor or activator B modeled by a Hill function $F(B)$. Show that in the presence of transcriptional delay τ^m, the dynamics of the active mRNA can be written as

$$\frac{dm^*(t)}{dt} = e^{-\tau^m} F(B(t - \tau^m)) - \bar{\delta} m^*.$$

2.8 Derive the expression of the parameters α, α_0 and K for the Hill function given in equation (2.20), which is the form obtained for transcriptional repression with a leaky repressor.

2.9 Consider the form of the Hill function in the presence of an activator with some basal level of expression given in equation (2.22). Derive the expressions of α, K and α_0,

2.10 Derive the form of the Hill functions for combinatorial promoters with leakiness given in expressions (2.26)–(2.27).

2.11 Consider the phosphorylation reactions described in Section 2.4, but suppose that the kinase concentration Z is not constant, but is instead produced and decays according to the reaction $Z \underset{k(t)}{\overset{\gamma}{\rightleftharpoons}} \emptyset$. How should the system in equation (2.29) be modified? Use a MATLAB simulation to apply a periodic input stimulus $k(t)$ using parameter values: $k_1 = k_2 = 1 \ \text{min}^{-1}$, $a_1 = a_2 = 10 \ \text{nM}^{-1} \ \text{min}^{-1}$, $d_1 = d_2 = 10 \ \text{min}^{-1}$, $\gamma = 0.01 \ \text{min}^{-1}$. Is the cycle capable of "tracking" the input stimulus? If yes, to what extent when the frequency of $k(t)$ is increased? What are the tracking properties depending on?

2.12 Another model for the phosphorylation reactions, referred to as *one-step reaction model*, is given by $Z + X \longrightarrow X^* + Z$ and $Y + X^* \longrightarrow X + Y$, in which the complex formations are neglected. Write down the differential equation model and compare the differential equation of X^* to that of equation (2.29). List the assumptions under which the one-step reaction model is a good approximation of the two-step reaction model.

2.13 (Competitive inhibition) Derive the expression of the production rate of X^* in a phosphorylation cycle in the presence of a competitive inhibitor I for Z.

2.14 (Non-absolute activator) Derive the expression of the production rate of X^* in a phosphorylation cycle in the presence of a non-absolute activator A for Z.

2.15 Consider the model of phosphotransfer systems of equation (2.32) and determine how the steady state transfer curve changes when the values of k_3 and k_4 are changed.

Chapter 3
Analysis of Dynamic Behavior

In this chapter, we describe some of the tools from dynamical systems and feedback control theory that will be used in the rest of the text to analyze and design biological circuits. We focus here on deterministic models and the associated analyses; stochastic methods are given in Chapter 4.

3.1 Analysis near equilibria

As in the case of many other classes of dynamical systems, a great deal of insight into the behavior of a biological system can be obtained by analyzing the dynamics of the system subject to small perturbations around a known solution. We begin by considering the dynamics of the system near an equilibrium point, which is one of the simplest cases and provides a rich set of methods and tools.

In this section we will model the dynamics of our system using the input/output modeling formalism described in Chapter 1:

$$\frac{dx}{dt} = f(x, \theta, u), \qquad y = h(x, \theta), \tag{3.1}$$

where $x \in \mathbb{R}^n$ is the system state, $\theta \in \mathbb{R}^p$ are the system parameters and $u \in \mathbb{R}^q$ is a set of external inputs (including disturbances and noise). The system state x is a vector whose components will represent concentration of species, such as transcription factors, enzymes, substrates and DNA promoter sites. The system parameters θ are also represented as a vector, whose components will represent biochemical parameters such as association and dissociation rate constants, production rate constants, decay rate constants and dissociation constants. The input u is a vector whose components will represent concentration of a number of possible physical entities, including kinases, allosteric effectors and some transcription factors. The output $y \in \mathbb{R}^m$ of the system represents quantities that can be measured or that are of interest for the specific problem under study.

Example 3.1 (Transcriptional component). Consider a promoter controlling a gene g that can be regulated by a transcription factor Z. Let m and G represent the mRNA and protein expressed by gene g. We can view this as a system in which $u = Z$ is the concentration of transcription factor regulating the promoter, the state $x = (x_1, x_2)$ is such that $x_1 = m$ is the concentration of mRNA and $x_2 = G$ is the concentration of protein, which we can take as the output of interest, that is, $y = G = x_2$. Assuming that the transcription factor regulating the promoter is a repressor, the system dynamics can be described by the following system:

$$\frac{dx_1}{dt} = \frac{\alpha}{1 + (u/K)^n} - \delta x_1, \qquad \frac{dx_2}{dt} = \kappa x_1 - \gamma x_2, \qquad y = x_2, \qquad (3.2)$$

in which $\theta = (\alpha, K, \delta, \kappa, \gamma, n)$ is the vector of system parameters. In this case, we have that

$$f(x, \theta, u) = \begin{pmatrix} \dfrac{\alpha}{1 + (u/K)^n} - \delta x_1 \\ \kappa x_1 - \gamma x_2 \end{pmatrix}, \qquad h(x, \theta) = x_2.$$

$$\nabla$$

Note that we have chosen to explicitly model the system parameters θ, which can be thought of as an additional set of (mainly constant) inputs to the system.

Equilibrium points and stability [1]

We begin by considering the case where the input u and parameters θ in equation (3.1) are fixed and hence we can write the dynamics of the system as

$$\frac{dx}{dt} = f(x). \qquad (3.3)$$

An *equilibrium point* of the dynamical system represents a stationary condition for the dynamics. We say that a state x_e is an equilibrium point for a dynamical system if $f(x_e) = 0$. If a dynamical system has an initial condition $x(0) = x_e$, then it will stay at the equilibrium point: $x(t) = x_e$ for all $t \geq 0$.

Equilibrium points are one of the most important features of a dynamical system since they define the states corresponding to constant operating conditions. A dynamical system can have zero, one or more equilibrium points.

The *stability* of an equilibrium point determines whether or not solutions nearby the equilibrium point remain close, get closer or move further away. An equilibrium point x_e is *stable* if solutions that start near x_e stay close to x_e. Formally, we say that the equilibrium point x_e is stable if for all $\epsilon > 0$, there exists a $\delta > 0$ such that

$$\|x(0) - x_e\| < \delta \quad \implies \quad \|x(t) - x_e\| < \epsilon \quad \text{for all } t > 0,$$

where $x(t)$ represents the solution to the differential equation (3.3) with initial condition $x(0)$. Note that this definition does not imply that $x(t)$ approaches x_e as time increases but just that it stays nearby. Furthermore, the value of δ may depend on ϵ, so that if we wish to stay very close to the solution, we may have to start very, very

---

[1]The material of this section is adopted from Åström and Murray [1].

close ($\delta \ll \epsilon$). This type of stability is also called *stability in the sense of Lyapunov*. If an equilibrium point is stable in this sense and the trajectories do not converge, we say that the equilibrium point is *neutrally stable*.

An example of a neutrally stable equilibrium point is shown in Figure 3.1. The

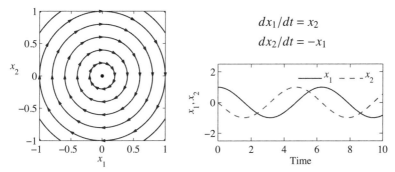

Figure 3.1: Phase portrait (trajectories in the state space) on the left and time domain simulation on the right for a system with a single stable equilibrium point. The equilibrium point x_e at the origin is stable since all trajectories that start near x_e stay near x_e.

figure shows the set of state trajectories starting at different initial conditions in the two-dimensional state space, also called the *phase plane*. From this set, called the *phase portrait*, we see that if we start near the equilibrium point, then we stay near the equilibrium point. Indeed, for this example, given any ϵ that defines the range of possible initial conditions, we can simply choose $\delta = \epsilon$ to satisfy the definition of stability since the trajectories are perfect circles.

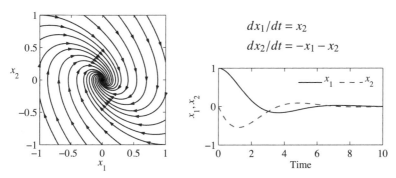

Figure 3.2: Phase portrait and time domain simulation for a system with a single asymptotically stable equilibrium point. The equilibrium point x_e at the origin is asymptotically stable since the trajectories converge to this point as $t \to \infty$.

An equilibrium point x_e is *asymptotically stable* if it is stable in the sense of Lyapunov and also $x(t) \to x_e$ as $t \to \infty$ for $x(0)$ sufficiently close to x_e. This corresponds to the case where all nearby trajectories converge to the stable solution for large time. Figure 3.2 shows an example of an asymptotically stable equilibrium point. Note from the phase portraits that not only do all trajectories stay near the equilibrium point at the origin, but that they also all approach the origin as t gets large (the directions of the arrows on the phase portrait show the direction in which the trajectories move).

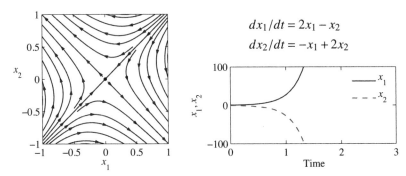

Figure 3.3: Phase portrait and time domain simulation for a system with a single unstable equilibrium point. The equilibrium point x_e at the origin is unstable since not all trajectories that start near x_e stay near x_e. The sample trajectory on the right shows that the trajectories very quickly depart from zero.

An equilibrium point x_e is *unstable* if it is not stable. More specifically, we say that an equilibrium point x_e is unstable if given some $\epsilon > 0$, there does *not* exist a $\delta > 0$ such that if $\|x(0) - x_e\| < \delta$, then $\|x(t) - x_e\| < \epsilon$ for all t. An example of an unstable equilibrium point is shown in Figure 3.3.

The definitions above are given without careful description of their domain of applicability. More formally, we define an equilibrium point to be *locally stable* (or *locally asymptotically stable*) if it is stable for all initial conditions $x \in B_r(a)$, where

$$B_r(a) = \{x : \|x - a\| < r\}$$

is a ball of radius r around a and $r > 0$. A system is *globally stable* if it is stable for all $r > 0$. Systems whose equilibrium points are only locally stable can have interesting behavior away from equilibrium points (see [1], Section 4.4).

To better understand the dynamics of the system, we can examine the set of all initial conditions that converge to a given asymptotically stable equilibrium point. This set is called the *region of attraction* for the equilibrium point. In general, computing regions of attraction is difficult. However, even if we cannot determine the region of attraction, we can often obtain patches around the stable equilibria that are attracting. This gives partial information about the behavior of the system.

For planar dynamical systems, equilibrium points have been assigned names based on their stability type. An asymptotically stable equilibrium point is called a *sink* or sometimes an *attractor*. An unstable equilibrium point can be either a *source*, if all trajectories lead away from the equilibrium point, or a *saddle*, if some trajectories lead to the equilibrium point and others move away (this is the situ-

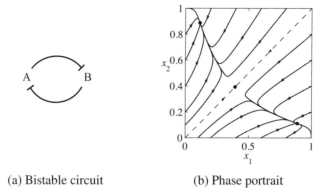

(a) Bistable circuit   (b) Phase portrait

Figure 3.4: (a) Diagram of a bistable gene circuit composed of two genes. (b) Phase portrait showing the trajectories converging to either one of the two possible stable equilibria depending on the initial condition. The parameters are $\beta_1 = \beta_2 = 1$ μM/min, $K_1 = K_2 = 0.1$ μM, and $\gamma = 1$ min^{-1}.

ation pictured in Figure 3.3). Finally, an equilibrium point that is stable but not asymptotically stable (i.e., neutrally stable, such as the one in Figure 3.1) is called a *center*.

Example 3.2 (Bistable gene circuit). Consider a system composed of two genes that express transcription factors repressing each other as shown in Figure 3.4a. Denoting the concentration of protein A by x_1 and that of protein B by x_2, and neglecting the mRNA dynamics, the system can be modeled by the following differential equations:

$$\frac{dx_1}{dt} = \frac{\beta_1}{1 + (x_2/K_2)^n} - \gamma x_1, \qquad \frac{dx_2}{dt} = \frac{\beta_2}{1 + (x_1/K_1)^n} - \gamma x_2.$$

Figure 3.4b shows the phase portrait of the system. This system is bistable because there are two (asymptotically) stable equilibria. Specifically, the trajectories converge to either of two possible equilibria: one where x_1 is high and x_2 is low and the other where x_1 is low and x_2 is high. A trajectory will approach the first equilibrium point if the initial condition is below the dashed line, called the separatrix, while it will approach the second one if the initial condition is above the separatrix. Hence, the region of attraction of the first equilibrium is the region of the plane below the separatrix and the region of attraction of the second one is the portion of the plane above the separatrix.   ▽

Nullcline analysis

Nullcline analysis is a simple and intuitive way to determine the stability of an equilibrium point for systems in $\mathbb{R}^2$. Consider the system with $x = (x_1, x_2) \in \mathbb{R}^2$

described by the differential equations

$$\frac{dx_1}{dt} = f_1(x_1, x_2), \qquad \frac{dx_2}{dt} = f_2(x_1, x_2).$$

The nullclines of this system are given by the two curves in the x_1, x_2 plane in which $f_1(x_1, x_2) = 0$ and $f_2(x_1, x_2) = 0$. The nullclines intersect at the equilibria of the system x_e. Figure 3.5 shows an example in which there is a unique equilibrium.

The stability of the equilibrium is deduced by inspecting the direction of the trajectory of the system starting at initial conditions x close to the equilibrium x_e. The direction of the trajectory can be obtained by determining the signs of f_1 and f_2 in each of the regions in which the nullclines partition the plane around the equilibrium x_e. If $f_1 < 0$ ($f_1 > 0$), we have that x_1 is going to decrease (increase) and similarly if $f_2 < 0$ ($f_2 > 0$), we have that x_2 is going to decrease (increase). In Figure 3.5, we show a case in which $f_1 < 0$ on the right-hand side of the nullcline $f_1 = 0$ and $f_1 > 0$ on the left-hand side of the same nullcline. Similarly, we have chosen a case in which $f_2 < 0$ above the nullcline $f_2 = 0$ and $f_2 > 0$ below the same nullcline. Given these signs, it is clear from the figure that starting from any point x close to x_e the vector field will always point toward the equilibrium x_e and hence the trajectory will tend toward such equilibrium. In this case, it then follows that the equilibrium x_e is asymptotically stable.

Example 3.3 (Negative autoregulation). As an example, consider expression of a gene with negative feedback. Let x_1 represent the mRNA concentration and x_2 represent the protein concentration. Then, a simple model (in which for simplicity we have assumed all parameters to be 1) is given by

$$\frac{dx_1}{dt} = \frac{1}{1 + x_2} - x_1, \qquad \frac{dx_2}{dt} = x_1 - x_2,$$

so that $f_1(x_1, x_2) = 1/(1 + x_2) - x_1$ and $f_2(x_1, x_2) = x_1 - x_2$. Figure 3.5a exactly represents the situation for this example. In fact, we have that

$$f_1(x_1, x_2) < 0 \iff x_1 > \frac{1}{1 + x_2}, \qquad f_2(x_1, x_2) < 0 \iff x_2 > x_1,$$

which provides the direction of the vector field as shown in Figure 3.5a. As a consequence, the equilibrium point is stable. The phase portrait of Figure 3.5b confirms the fact since the trajectories all converge to the unique equilibrium point.

$$\nabla$$

Stability analysis via linearization

For systems with more than two states, the graphical technique of nullcline analysis cannot be used. Hence, we must resort to other techniques to determine stability.

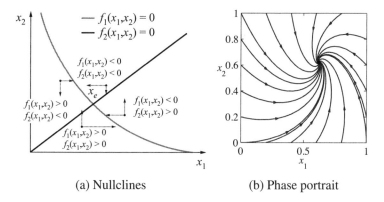

(a) Nullclines   (b) Phase portrait

Figure 3.5: (a) Example of nullclines for a system with a single equilibrium point x_e. To understand the stability of the equilibrium point x_e, one traces the direction of the vector field (f_1, f_2) in each of the four regions in which the nullclines partition the plane. If in each region the vector field points toward the equilibrium point, then such a point is asymptotically stable. (b) Phase portrait for the negative autoregulation example.

Consider a linear dynamical system of the form

$$\frac{dx}{dt} = Ax, \quad x(0) = x_0, \tag{3.4}$$

where $A \in \mathbb{R}^{n \times n}$. For a linear system, the stability of the equilibrium at the origin can be determined from the eigenvalues of the matrix A:

$$\lambda(A) = \{s \in \mathbb{C} : \det(sI - A) = 0\}.$$

The polynomial $\det(sI - A)$ is the *characteristic polynomial* and the eigenvalues are its roots. We use the notation λ_j for the jth eigenvalue of A and $\lambda(A)$ for the set of all eigenvalues of A, so that $\lambda_j \in \lambda(A)$. For each eigenvalue λ_j there is a corresponding eigenvector $v_j \in \mathbb{C}^n$, which satisfies the equation $Av_j = \lambda_j v_j$.

In general λ can be complex-valued, although if A is real-valued, then for any eigenvalue λ, its complex conjugate λ^* will also be an eigenvalue. The origin is always an equilibrium point for a linear system. Since the stability of a linear system depends only on the matrix A, we find that stability is a property of the system. For a linear system we can therefore talk about the stability of the system rather than the stability of a particular solution or equilibrium point.

The easiest class of linear systems to analyze are those whose system matrices are in diagonal form. In this case, the dynamics have the form

$$\frac{dx}{dt} = \begin{pmatrix} \lambda_1 & & & 0 \\ & \lambda_2 & & \\ & & \ddots & \\ 0 & & & \lambda_n \end{pmatrix} x. \tag{3.5}$$

It is easy to see that the state trajectories for this system are independent of each other, so that we can write the solution in terms of n individual systems $\dot{x}_j = \lambda_j x_j$. Each of these scalar solutions is of the form

$$x_j(t) = e^{\lambda_j t} x_j(0).$$

We see that the equilibrium point $x_e = 0$ is stable if $\lambda_j \leq 0$ and asymptotically stable if $\lambda_j < 0$.

Another simple case is when the dynamics are in the block diagonal form

$$\frac{dx}{dt} = \begin{pmatrix} \sigma_1 & \omega_1 & & 0 & 0 \\ -\omega_1 & \sigma_1 & & 0 & 0 \\ 0 & 0 & \ddots & \vdots & \vdots \\ 0 & 0 & & \sigma_m & \omega_m \\ 0 & 0 & & -\omega_m & \sigma_m \end{pmatrix} x.$$

In this case, the eigenvalues can be shown to be $\lambda_j = \sigma_j \pm i\omega_j$. We once again can separate the state trajectories into independent solutions for each pair of states, and the solutions are of the form

$$x_{2j-1}(t) = e^{\sigma_j t}(x_{2j-1}(0)\cos\omega_j t + x_{2j}(0)\sin\omega_j t),$$
$$x_{2j}(t) = e^{\sigma_j t}(-x_{2j-1}(0)\sin\omega_j t + x_{2j}(0)\cos\omega_j t),$$

where $j = 1, 2, \ldots, m$. We see that this system is asymptotically stable if and only if $\sigma_j = \operatorname{Re}\lambda_j < 0$. It is also possible to combine real and complex eigenvalues in (block) diagonal form, resulting in a mixture of solutions of the two types.

Very few systems are in one of the diagonal forms above, but some systems can be transformed into these forms via coordinate transformations. One such class of systems is those for which the A matrix has distinct (non-repeating) eigenvalues. In this case there is a matrix $T \in \mathbb{R}^{n \times n}$ such that the matrix TAT^{-1} is in (block) diagonal form, with the block diagonal elements corresponding to the eigenvalues of the original matrix A. If we choose new coordinates $z = Tx$, then

$$\frac{dz}{dt} = T\dot{x} = TAx = TAT^{-1}z$$

and the linear system has a (block) diagonal A matrix. Furthermore, the eigenvalues of the transformed system are the same as the original system since if v is an eigenvector of A, then $w = Tv$ can be shown to be an eigenvector of TAT^{-1}. We can reason about the stability of the original system by noting that $x(t) = T^{-1}z(t)$, and so if the transformed system is stable (or asymptotically stable), then the original system has the same type of stability.

This analysis shows that for linear systems with distinct eigenvalues, the stability of the system can be completely determined by examining the real part of the eigenvalues of the dynamics matrix. For more general systems, we make use of the following theorem, proved in [1]:

Theorem 3.1 (Stability of a linear system). *The system*

$$\frac{dx}{dt} = Ax$$

is asymptotically stable if and only if all eigenvalues of A all have a strictly negative real part and is unstable if any eigenvalue of A has a strictly positive real part.

In the case in which the system state is two-dimensional, that is, $x \in \mathbb{R}^2$, we have a simple way of determining the eigenvalues of a matrix A. Specifically, denote by $\text{tr}(A)$ the trace of A, that is, the sum of the diagonal terms, and let $\det(A)$ be the determinant of A. Then, we have that the two eigenvalues are given by

$$\lambda_{1,2} = \frac{1}{2}\left(\text{tr}(A) \pm \sqrt{\text{tr}(A)^2 - 4\det(A)}\right).$$

Both eigenvalues have negative real parts when (i) $\text{tr}(A) < 0$ and (ii) $\det(A) > 0$.

An important feature of differential equations is that it is often possible to determine the local stability of an equilibrium point by approximating the system by a linear system. Suppose that we have a nonlinear system

$$\frac{dx}{dt} = f(x)$$

that has an equilibrium point at x_e. Computing the Taylor series expansion of the vector field, we can write

$$\frac{dx}{dt} = f(x_e) + \left.\frac{\partial f}{\partial x}\right|_{x_e} (x - x_e) + \text{higher-order terms in } (x - x_e).$$

Since $f(x_e) = 0$, we can approximate the system by choosing a new state variable $z = x - x_e$ and writing

$$\frac{dz}{dt} = Az, \qquad \text{where} \quad A = \left.\frac{\partial f}{\partial x}\right|_{x_e}. \tag{3.6}$$

We call the system (3.6) the *linear approximation* of the original nonlinear system or the *linearization* at x_e. We also refer to matrix A as the *Jacobian matrix* of the original nonlinear system.

The fact that a linear model can be used to study the behavior of a nonlinear system near an equilibrium point is a powerful one. Indeed, we can take this even further and use a local linear approximation of a nonlinear system to design a feedback law that keeps the system near its equilibrium point (design of dynamics). Thus, feedback can be used to make sure that solutions remain close to the equilibrium point, which in turn ensures that the linear approximation used to stabilize it is valid.

Example 3.4 (Negative autoregulation). Consider again the negatively autoregulated gene modeled by the equations

$$\frac{dx_1}{dt} = \frac{1}{1 + x_2} - x_1, \qquad \frac{dx_2}{dt} = x_1 - x_2.$$

In this case,

$$f(x) = \begin{pmatrix} \frac{1}{1+x_2} - x_1 \\ x_1 - x_2 \end{pmatrix},$$

so that, letting $x_e = (x_{1,e}, x_{2,e})$, the Jacobian matrix is given by

$$A = \left. \frac{\partial f}{\partial x} \right|_{x_e} = \begin{pmatrix} -1 & -\frac{1}{(1+x_{2,e})^2} \\ 1 & -1 \end{pmatrix}.$$

It follows that $\text{tr}(A) = -2 < 0$ and that $\det(A) = 1 + 1/(1 + x_{2,e})^2 > 0$. Hence, independently of the value of the equilibrium point, the eigenvalues both have negative real parts, which implies that the equilibrium point x_e is asymptotically stable. $\quad \nabla$

Frequency domain analysis

Frequency domain analysis is a way to understand how well a system can respond to rapidly changing input stimuli. As a general rule, most physical systems display an increased difficulty in responding to input stimuli as the frequency of variation increases: when the input stimulus changes faster than the natural time scales of the system, the system becomes incapable of responding. If instead the input is changing much more slowly than the natural time scales of the system, the system will have enough time to respond to the input. That is, the system behaves like a "low-pass filter." The cut-off frequency at which the system does not display a significant response is called the *bandwidth* and quantifies the dominant time scale. To identify this dominant time scale, we can perform input/output experiments in which the system is excited with periodic inputs at various frequencies. Then, we can plot the amplitude of response of the output as a function of the frequency of the input stimulation to obtain the "frequency response" of the system.

Example 3.5 (Phosphorylation cycle). To illustrate the basic ideas, we consider the frequency response of a phosphorylation cycle, in which enzymatic reactions are each modeled by a one-step reaction. Referring to Figure 3.6a, we have that the one-step reactions involved are given by

$$Z + X \xrightarrow{k_1} Z + X^*, \qquad Y + X^* \xrightarrow{k_2} Y + X,$$

with conservation law $X + X^* = X_{\text{tot}}$. Let Y_{tot} be the total amount of phosphatase. We assume that the kinase Z has a time-varying concentration, which we view as the *input* to the system, while X^* is the *output* of the system.

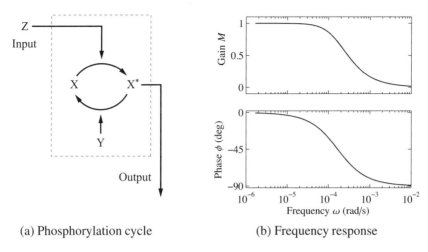

(a) Phosphorylation cycle   (b) Frequency response

Figure 3.6: (a) Diagram of a phosphorylation cycle, in which Z is the kinase, X is the substrate, and Y is the phosphatase. (b) Bode plot showing the magnitude M and phase lag ϕ for the frequency response of a one-step reaction model of the phosphorylation system on the left. The parameters are $\beta = \gamma = 0.01 \text{ min}^{-1}$.

The differential equation model for the dynamics is given by

$$\frac{dX^*}{dt} = k_1 Z(t)(X_{\text{tot}} - X^*) - k_2 Y_{\text{tot}} X^*.$$

If we assume that the cycle is weakly activated ($X^* \ll X_{\text{tot}}$), the above equation is well approximated by

$$\frac{dX^*}{dt} = \beta Z(t) - \gamma X^*, \tag{3.7}$$

where $\beta = k_1 X_{\text{tot}}$ and $\gamma = k_2 Y_{\text{tot}}$. To determine the frequency response, we set the input $Z(t)$ to a periodic function. It is customary to take sinusoidal functions as the input signal as they lead to an easy way to calculate the frequency response. Let then $Z(t) = A_0 \sin(\omega t)$.

Since equation (3.7) is linear in the state X^* and input Z, it can be directly integrated to yield

$$X^*(t) = \frac{A_0 \beta}{\sqrt{\omega^2 + \gamma^2}} \sin(\omega t - \tan^{-1}(\omega/\gamma)) - \frac{A_0 \beta \omega}{(\omega^2 + \gamma^2)} e^{-\gamma t}.$$

The second term dies out for t large enough. Hence, the steady state response is given by the first term. In particular, the amplitude of response is given by $A_0 \beta / \sqrt{\omega^2 + \gamma^2}$, in which the gain $\beta / \sqrt{\omega^2 + \gamma^2}$ depends both on the system parameters and on the frequency of the input stimulation. As the frequency of the input stimulation ω increases, the amplitude of the response decreases and approaches zero for very high frequencies. Also, the argument of the sine function shows a negative phase shift of $\tan^{-1}(\omega/\gamma)$, which indicates that there is an increased lag in responding to the input when the frequency increases. Hence, the key quantities in the frequency response are the magnitude $M(\omega)$, also called *gain* of the system, and phase lag $\phi(\omega)$ given by

$$M(\omega) = \frac{\beta}{\sqrt{\omega^2 + \gamma^2}}, \qquad \phi(\omega) = -\tan^{-1}\left(\frac{\omega}{\gamma}\right).$$

These are plotted in Figure 3.6b, a type of figure known as a *Bode plot*.

The bandwidth of the system, denoted ω_B, is the frequency at which the gain drops below $M(0)/\sqrt{2}$. In this case, the bandwidth is given by $\omega_B = \gamma = k_2 Y_{\text{tot}}$, which implies that the bandwidth of the system can be made larger by increasing the amount of phosphatase. However, note that since $M(0) = \beta/\gamma = k_1 X_{\text{tot}}/(k_2 Y_{\text{tot}})$, increased phosphatase will also result in decreased amplitude of response. Hence, if we want to increase the bandwidth of the system while keeping the value of $M(0)$ (also called the *zero frequency gain*) unchanged, one should increase the total amounts of substrate and phosphatase in comparable proportions. Fixing the value of the zero frequency gain, the bandwidth of the system increases with increased amounts of phosphatase and substrate. ∇

More generally, the *frequency response* of a linear system with one input and one output

$$\dot{x} = Ax + Bu, \qquad y = Cx + Du$$

is the response of the system to a sinusoidal input $u = a \sin \omega t$ with input amplitude a and frequency ω. The *transfer function* for a linear system is given by

$$G_{yu}(s) = C(sI - A)^{-1}B + D$$

and represents the steady state response of a system to an exponential signal of the form $u(t) = e^{st}$ where $s \in \mathbb{C}$. In particular, the response to a sinusoid $u = a \sin \omega t$ is given by $y = Ma \sin(\omega t + \phi)$ where the gain M and phase lag ϕ can be determined from the transfer function evaluated at $s = i\omega$:

$$G_{yu}(i\omega) = Me^{i\phi}, \qquad \begin{aligned} M(\omega) &= |G_{yu}(i\omega)| = \sqrt{\text{Im}(G_{yu}(i\omega))^2 + \text{Re}(G_{yu}(i\omega))^2}, \\ \phi(\omega) &= \tan^{-1}\left(\frac{\text{Im}(G_{yu}(i\omega))}{\text{Re}(G_{yu}(i\omega))}\right), \end{aligned}$$

where $\text{Re}(\cdot)$ and $\text{Im}(\cdot)$ represent the real and imaginary parts of a complex number. For finite dimensional linear (or linearized) systems, the transfer function can be written as a ratio of polynomials in s:

$$G(s) = \frac{b(s)}{a(s)}.$$

The values of s at which the numerator vanishes are called the *zeros* of the transfer function and the values of s at which the denominator vanishes are called the *poles*.

The transfer function representation of an input/output linear system is essentially equivalent to the state space description, but we reason about the dynamics by looking at the transfer function instead of the state space matrices. For example, it can be shown that the poles of a transfer function correspond to the eigenvalues of the matrix A, and hence the poles determine the stability of the system. In addition, interconnections between subsystems often have simple representations in terms of transfer functions. For example, two systems G_1 and G_2 in series (with the output of the first connected to the input of the second) have a combined transfer function $G_{series}(s) = G_1(s)G_2(s)$, and two systems in parallel (a single input goes to both systems and the outputs are summed) has the transfer function $G_{parallel}(s) = G_1(s) + G_2(s)$.

Transfer functions are useful representations of linear systems because the properties of the transfer function can be related to the properties of the dynamics. In particular, the shape of the frequency response describes how the system responds to inputs and disturbances, as well as allows us to reason about the stability of interconnected systems. The Bode plot of a transfer function gives the magnitude and phase of the frequency response as a function of frequency and the *Nyquist plot* can be used to reason about stability of a closed loop system from the open loop frequency response ([1], Section 9.2).

Returning to our analysis of biomolecular systems, suppose we have a system whose dynamics can be written as

$$\dot{x} = f(x, \theta, u)$$

and we wish to understand how the solutions of the system depend on the parameters θ and input disturbances u. We focus on the case of an equilibrium solution $x(t; x_0, \theta_0) = x_e$. Let $z = x - x_e$, $\tilde{u} = u - u_0$ and $\tilde{\theta} = \theta - \theta_0$ represent the deviation of the state, input and parameters from their nominal values. Linearization can be performed in a way similar to the way it was performed for a system with no inputs. Specifically, we can write the dynamics of the perturbed system using its linearization as

$$\frac{dz}{dt} = \left(\frac{\partial f}{\partial x}\right)_{(x_e, \theta_0, u_0)} \cdot z \quad + \quad \left(\frac{\partial f}{\partial \theta}\right)_{(x_e, \theta_0, u_0)} \cdot \tilde{\theta} \quad + \quad \left(\frac{\partial f}{\partial u}\right)_{(x_e, \theta_0, u_0)} \cdot \tilde{u}.$$

This linear system describes small deviations from $x_e(\theta_0, u_0)$ but allows $\tilde{\theta}$ and $\tilde{u}$ to be time-varying instead of the constant case considered earlier.

To analyze the resulting deviations, it is convenient to look at the system in the frequency domain. Let $y = Cx$ be a set of values of interest. The transfer functions between $\tilde{\theta}$, $\tilde{u}$ and y are given by

$$G_{y\tilde{\theta}}(s) = C(sI - A)^{-1}B_\theta, \qquad G_{y\tilde{u}}(s) = C(sI - A)^{-1}B_u,$$

where

$$A = \frac{\partial f}{\partial x}\bigg|_{(x_e,\theta_0,u_0)}, \qquad B_\theta = \frac{\partial f}{\partial \theta}\bigg|_{(x_e,\theta_0,u_0)}, \qquad B_u = \frac{\partial f}{\partial u}\bigg|_{(x_e,\theta_0,u_0)}.$$

Note that if we let $s = 0$, we get the response to small, constant changes in parameters. For example, the change in the outputs y as a function of constant changes in the parameters is given by

$$G_{y\tilde\theta}(0) = -CA^{-1}B_\theta.$$

Example 3.6 (Transcriptional regulation). Consider a genetic circuit consisting of a single gene. The dynamics of the system are given by

$$\frac{dm}{dt} = F(P) - \delta m, \qquad \frac{dP}{dt} = \kappa m - \gamma P,$$

where m is the mRNA concentration and P is the protein concentration. Suppose that the mRNA degradation rate δ can change as a function of time and that we wish to understand the sensitivity with respect to this (time-varying) parameter. Linearizing the dynamics around the equilibrium point (m_e, P_e) corresponding to a nominal value δ_0 of the mRNA degradation rate, we obtain

$$A = \begin{pmatrix} -\delta_0 & F'(P_e) \\ \kappa & -\gamma \end{pmatrix}, \qquad B_\delta = \begin{pmatrix} -m_e \\ 0 \end{pmatrix}. \tag{3.8}$$

For the case of no feedback we have $F(P) = \alpha$ and $F'(P) = 0$, and the system has the equilibrium point at $m_e = \alpha/\delta_0$, $P_e = \kappa\alpha/(\gamma\delta_0)$. The transfer function from δ to P, after linearization about the steady state, is given by

$$G_{P\delta}^{\text{ol}}(s) = \frac{-\kappa m_e}{(s+\delta_0)(s+\gamma)},$$

where "ol" stands for open loop. For the case of negative regulation, we have

$$F(P) = \frac{\alpha}{1 + (P/K)^n} + \alpha_0,$$

and the resulting transfer function is given by

$$G_{P\delta}^{\text{cl}}(s) = \frac{\kappa m_e}{(s+\delta_0)(s+\gamma) + \kappa\sigma}, \qquad \sigma = -F'(P_e) = \frac{n\alpha P_e^{n-1}/K^n}{(1 + P_e^n/K^n)^2},$$

where "cl" stands for closed loop.

Figure 3.7 shows the frequency response for the two circuits. To make a meaningful comparison between open loop and closed loop systems, we select the parameters of the open loop system such that the equilibrium point for both open loop and closed loop systems are the same. This can be guaranteed if in the open loop system we choose, for example, $\alpha = P_e\delta_0/(\kappa/\gamma)$, in which P_e is the equilibrium value of P in the closed loop system. We see that the feedback circuit attenuates the response of the system to perturbations with low-frequency content but slightly amplifies perturbations at high frequency (compared to the open loop system). ∇

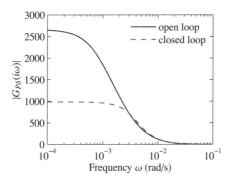

Figure 3.7: Attenuation of perturbations in a genetic circuit with linearization given by equation (3.8). The parameters of the closed loop system are given by $\alpha = 800$ µM/s, $\alpha_0 = 5 \times 10^{-4}$ µM/s, $\gamma = 0.001$ s^{-1}, $\delta_0 = 0.005$ s^{-1}, $\kappa = 0.02$ s^{-1}, $n = 2$, and $K = 0.025$ µM. For the open loop system, we have set $\alpha = P_e \delta_0 / (\kappa/\gamma)$ to make the steady state values of open loop and closed loop systems the same.

3.2 Robustness

The term "robustness" refers to the general ability of a system to continue to function in the presence of uncertainty. In the context of this text, we will want to be more precise. We say that a given function (of the circuit) is robust with respect to a set of specified perturbations if the sensitivity of that function to perturbations is small. Thus, to study robustness, we must specify both the function we are interested in and the set of perturbations that we wish to consider.

In this section we study the robustness of the system

$$\frac{dx}{dt} = f(x, \theta, u), \qquad y = h(x, \theta)$$

to various perturbations in the parameters θ and disturbance inputs u. The function we are interested in is modeled by the outputs y and hence we seek to understand how y changes if the parameters θ are changed by a small amount or if external disturbances u are present. We say that a system is robust with respect to these perturbations if y undergoes little change as these perturbations are introduced.

Parametric uncertainty

In addition to studying the input/output transfer curve and the stability of a given equilibrium point, we can also study how these features change with respect to changes in the system parameters θ. Let $y_e(\theta_0, u_0)$ represent the output corresponding to an equilibrium point x_e with fixed parameters θ_0 and external input u_0, so that $f(x_e, \theta_0, u_0) = 0$. We assume that the equilibrium point is stable and focus here on understanding how the value of the output, the location of the equilibrium point, and the dynamics near the equilibrium point vary as a function of changes in the parameters θ and external inputs u.

We start by assuming that $u = 0$ and investigate how x_e and y_e depend on θ; we will write $f(x, \theta)$ instead of $f(x, \theta, 0)$ to simplify notation. The simplest approach is to analytically solve the equation $f(x_e, \theta_0) = 0$ for x_e and then set $y_e = h(x_e, \theta_0)$. However, this is often difficult to do in closed form and so as an alternative we instead look at the linearized response given by

$$S_{x,\theta} := \frac{dx_e}{d\theta}\bigg|_{\theta_0}, \qquad S_{y,\theta} := \frac{dy_e}{d\theta}\bigg|_{\theta_0},$$

which are the (infinitesimal) changes in the equilibrium state and the output due to a change in the parameter. To determine $S_{x,\theta}$ we begin by differentiating the relationship $f(x_e(\theta), \theta) = 0$ with respect to θ:

$$\frac{df}{d\theta} = \frac{\partial f}{\partial x}\frac{dx_e}{d\theta} + \frac{\partial f}{\partial \theta} = 0 \implies S_{x,\theta} = \frac{dx_e}{d\theta} = -\left(\frac{\partial f}{\partial x}\right)^{-1}\frac{\partial f}{\partial \theta}\bigg|_{(x_e, \theta_0)}. \qquad (3.9)$$

Similarly, we can compute the output sensitivity as

$$S_{y,\theta} = \frac{dy_e}{d\theta} = \frac{\partial h}{\partial x}\frac{dx_e}{d\theta} + \frac{\partial h}{\partial \theta} = -\left(\frac{\partial h}{\partial x}\left(\frac{\partial f}{\partial x}\right)^{-1}\frac{\partial f}{\partial \theta} - \frac{\partial h}{\partial \theta}\right)\bigg|_{(x_e, \theta_0)}.$$

These quantities can be computed numerically and hence we can evaluate the effect of small (but constant) changes in the parameters θ on the equilibrium state x_e and corresponding output value y_e.

A similar analysis can be performed to determine the effects of small (but constant) changes in the external input u. Suppose that x_e depends on both θ and u, with $f(x_e, \theta_0, u_0) = 0$ and θ_0 and u_0 representing the nominal values. Then

$$\frac{dx_e}{d\theta}\bigg|_{(\theta_0, u_0)} = -\left(\frac{\partial f}{\partial x}\right)^{-1}\frac{\partial f}{\partial \theta}\bigg|_{(x_e, \theta_0, u_0)}, \qquad \frac{dx_e}{du}\bigg|_{(\theta_0, u_0)} = -\left(\frac{\partial f}{\partial x}\right)^{-1}\frac{\partial f}{\partial u}\bigg|_{(x_e, \theta_0, u_0)}.$$

The sensitivity matrix can be normalized by dividing the parameters by their nominal values and rescaling the outputs (or states) by their equilibrium values. If we define the scaling matrices

$$D^{x_e} = \text{diag}\{x_e\}, \qquad D^{y_e} = \text{diag}\{y_e\}, \qquad D^{\theta} = \text{diag}\{\theta\},$$

then the scaled sensitivity matrices can be written as

$$\bar{S}_{x,\theta} = (D^{x_e})^{-1}S_{x,\theta}D^{\theta}, \qquad \bar{S}_{y,\theta} = (D^{y_e})^{-1}S_{y,\theta}D^{\theta}. \qquad (3.10)$$

The entries in these matrices describe how a fractional change in a parameter gives a fractional change in the state or output, relative to the nominal values of the parameters and state or output.

Example 3.7 (Transcriptional regulation). Consider again the case of transcriptional regulation described in Example 3.6. We wish to study the response of the protein concentration to fluctuations in its parameters in two cases: a *constitutive promoter* (open loop) and self-repression (closed loop).

For the case of open loop we have $F(P) = \alpha$, and the system has the equilibrium point at $m_e = \alpha/\delta$, $P_e = \kappa\alpha/(\gamma\delta)$. The parameter vector can be taken as $\theta = (\alpha, \delta, \kappa, \gamma)$ and the state as $x = (m, P)$. Since we have a simple expression for the equilibrium concentrations, we can compute the sensitivity to the parameters directly:

$$\frac{\partial x_e}{\partial \theta} = \begin{pmatrix} \frac{1}{\delta} & -\frac{\alpha}{\delta^2} & 0 & 0 \\ \frac{\kappa}{\gamma\delta} & -\frac{\kappa\alpha}{\gamma\delta^2} & \frac{\alpha}{\gamma\delta} & -\frac{\kappa\alpha}{\delta\gamma^2} \end{pmatrix},$$

where the parameters are evaluated at their nominal values, but we leave off the subscript 0 on the individual parameters for simplicity. If we choose the parameters as $\theta_0 = (0.00138, 0.00578, 0.115, 0.00116)$, then the resulting sensitivity matrix evaluates to

$$S^{\text{open}}_{x_e, \theta} \approx \begin{pmatrix} 173 & -42 & 0 & 0 \\ 17300 & -4200 & 211 & -21100 \end{pmatrix}. \tag{3.11}$$

If we look instead at the scaled sensitivity matrix, then the open loop nature of the system yields a particularly simple form:

$$\bar{S}^{\text{open}}_{x_e, \theta} = \begin{pmatrix} 1 & -1 & 0 & 0 \\ 1 & -1 & 1 & -1 \end{pmatrix}. \tag{3.12}$$

In other words, a 10% change in any of the parameters will lead to a comparable positive or negative change in the equilibrium values.

For the case of negative regulation, we have

$$F(P) = \frac{\alpha}{1 + (P/K)^n} + \alpha_0,$$

and the equilibrium points satisfy

$$m_e = \frac{\gamma}{\kappa} P_e, \qquad \frac{\alpha}{1 + P_e^n/K^n} + \alpha_0 = \delta m_e = \frac{\delta\gamma}{\kappa} P_e. \tag{3.13}$$

In order to make a proper comparison with the previous case, we need to choose the parameters so that the equilibrium concentrations m_e, P_e match those of the open loop system. We can do this by modifying the promoter strength α and/or the RBS strength, which is proportional to κ, so that the second formula in equation (3.13) is satisfied or, equivalently, choose the parameters for the open loop case so that they match the closed loop steady state protein concentration (see Example 2.2).

Rather than attempt to solve for the equilibrium point in closed form, we instead investigate the sensitivity using the computations in equation (3.13). The state, dynamics and parameters are given by

$$x = \begin{pmatrix} m & P \end{pmatrix}, \qquad f(x, \theta) = \begin{pmatrix} F(P) - \delta m \\ \kappa m - \gamma P \end{pmatrix}, \qquad \theta = \begin{pmatrix} \alpha_0 & \delta & \kappa & \gamma & \alpha & n & K \end{pmatrix}.$$

Note that the parameters are ordered such that the first four parameters match the open loop system. The linearizations are given by

$$\frac{\partial f}{\partial x} = \begin{pmatrix} -\delta & F'(P_e) \\ \beta & -\gamma \end{pmatrix}, \qquad \frac{\partial f}{\partial \theta} = \begin{pmatrix} 1 & -m_e & 0 & 0 & \partial F/\partial \alpha & \partial F/\partial n & \partial F/\partial K \\ 0 & 0 & m_e & -P_e & 0 & 0 & 0 \end{pmatrix},$$

where again the parameters are taken to be at their nominal values and the derivatives are evaluated at the equilibrium point. From this we can compute the sensitivity matrix as

$$S_{x,\theta} = \begin{pmatrix} -\dfrac{\gamma}{\gamma\delta-\kappa F'} & \dfrac{\gamma m}{\gamma\delta-\kappa F'} & -\dfrac{mF'}{\gamma\delta-\kappa F'} & \dfrac{PF'}{\gamma\delta-\kappa F'} & -\dfrac{\gamma\partial F/\partial\alpha}{\gamma\delta-\kappa F'} & -\dfrac{\gamma\partial F/\partial n}{\gamma\delta-\kappa F'} & -\dfrac{\gamma\partial F/\partial K}{\gamma\delta-\kappa F'} \\ -\dfrac{\kappa}{\gamma\delta-\kappa F'} & \dfrac{\kappa m}{\gamma\delta-\kappa F'} & -\dfrac{\delta m}{\gamma\delta-\kappa F'} & \dfrac{\delta P}{\gamma\delta-\kappa F'} & -\dfrac{\kappa\partial F/\partial\alpha_1}{\gamma\delta-\kappa F'} & -\dfrac{\kappa\partial F/\partial n}{\gamma\delta-\kappa F'} & -\dfrac{\kappa\partial F/\partial K}{\gamma\delta-\kappa F'} \end{pmatrix},$$

where $F' = \partial F/\partial P$ and all other derivatives of F are evaluated at the nominal parameter values and the corresponding equilibrium point. In particular, we take nominal parameters as $\theta = (5 \cdot 10^{-4}, 0.005, 0.115, 0.001, 800, 2, 0.025)$.

We can now evaluate the sensitivity at the same protein concentration as we use in the open loop case. The equilibrium point is given by

$$x_e = \begin{pmatrix} m_e \\ P_e \end{pmatrix} = \begin{pmatrix} 0.239 \\ 23.9 \end{pmatrix}$$

and the sensitivity matrix is

$$S_{x_e,\theta}^{\text{closed}} \approx \begin{pmatrix} 76 & -18 & -1.15 & 115 & 0.00008 & -0.45 & 5.34 \\ 7611 & -1816 & 90 & -9080. & 0.008 & -45 & 534 \end{pmatrix}.$$

The scaled sensitivity matrix becomes

$$\bar{S}_{x_e,\theta}^{\text{closed}} \approx \begin{pmatrix} 0.159 & -0.44 & -0.56 & 0.56 & 0.28 & -3.84 & 0.56 \\ 0.159 & -0.44 & 0.44 & -0.44 & 0.28 & -3.84 & 0.56 \end{pmatrix}. \tag{3.14}$$

Comparing this equation with equation (3.12), we see that there is reduction in the sensitivity with respect to most parameters. In particular, we become less sensitive to those parameters that are not part of the feedback (columns 2–4), but there is higher sensitivity with respect to some of the parameters that are part of the feedback mechanism (particularly n).   ∇

More generally, we may wish to evaluate the sensitivity of a (non-constant) solution to parameter changes. This can be done by computing the function $dx(t)/d\theta$, which describes how the state changes at each instant in time as a function of (small) changes in the parameters θ. This can be used, for example, to understand how we can change the parameters to obtain a desired behavior or to determine the most critical parameters that determine a specific dynamical feature of the system under study.

Let $x(t, \theta_0)$ be a solution of the nominal system

$$\dot{x} = f(x, \theta_0, u), \qquad x(0) = x_0.$$

To compute $dx/d\theta$, we write a differential equation for how it evolves in time:

$$\frac{d}{dt}\left(\frac{dx}{d\theta}\right) = \frac{d}{d\theta}\left(\frac{dx}{dt}\right) = \frac{d}{d\theta}(f(x, \theta, u)) = \frac{\partial f}{\partial x}\frac{dx}{d\theta} + \frac{\partial f}{\partial \theta}.$$

This is a differential equation with $n \times m$ states given by the entries of the matrix $S_{x,\theta}(t) = dx(t)/d\theta$ and with initial condition $S_{x,\theta}(0) = 0$ (since changes to the parameters do not affect the initial conditions).

To solve these equations, we must simultaneously solve for the state x and the sensitivity $S_{x,\theta}$ (whose dynamics depend on x). Thus, letting

$$M(t, \theta_0) := \left.\frac{\partial f}{\partial x}(x, \theta, u)\right|_{x=x(t,\theta_0), \theta=\theta_0}, \qquad N(t, \theta_0) := \left.\frac{\partial f}{\partial \theta}(x, \theta, u)\right|_{x=x(t,\theta_0), \theta=\theta_0},$$

we solve the set of $n + nm$ coupled differential equations

$$\frac{dx}{dt} = f(x, \theta_0, u), \qquad \frac{dS_{x,\theta}}{dt} = M(t, \theta_0)S_{x,\theta} + N(t, \theta_0), \tag{3.15}$$

with initial condition $x(0) = x_0$ and $S_{x,\theta}(0) = 0$.

This differential equation generalizes our previous results by allowing us to evaluate the sensitivity around a (non-constant) trajectory. Note that in the special case in which we are at an equilibrium point and the dynamics for $S_{x,\theta}$ are stable, the steady state solution of equation (3.15) is identical to that obtained in equation (3.9). However, equation (3.15) is much more general, allowing us to determine the change in the state of the system at a fixed time T, for example. This equation also does not require that our solution stay near an equilibrium point; it only requires that our perturbations in the parameters are sufficiently small. An example of how to apply this equation to study the effect of parameter changes on an oscillator is given in Section 5.4.

Several simulation tools include the ability to do sensitivity analysis of this sort, including COPASI and the MATLAB SimBiology toolbox.

Adaptation and disturbance rejection

In this section, we study how systems can keep a desired output response even in the presence of external disturbances. This property is particularly important for biomolecular systems, which are usually subject to a wide range of perturbations. These perturbations or disturbances can represent a number of different physical entities, including changes in the circuit's cellular environment, unmodeled/undesired interactions with other biological circuits present in the cell, or parameters whose values are uncertain.

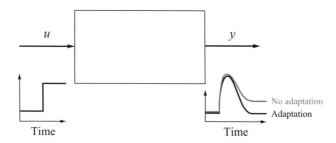

Figure 3.8: Adaptation property. The system is said to have the adaptation property if the steady state value of the output does not depend on the steady state value of the input. Hence, after a constant input perturbation, the output returns to its original value.

Here, we represent the disturbance input to the system of interest by u and we will say that the system adapts to the input u when the steady state value of its output y is independent of the (constant) nonzero value of the input (Figure 3.8). That is, the system's output is robust to the disturbance input. Basically, after the input changes to a constant nonzero value, the output returns to its original value after a transient perturbation. Adaptation corresponds to the concept of *disturbance rejection* in control theory. The full notion of disturbance rejection is more general, depends on the specific disturbance input and is often studied using the internal model principle [17].

We illustrate two main mechanisms to attain adaptation: integral feedback and incoherent feedforward loops (IFFLs). Here, we follow a similar treatment as that of [89]. In particular, we study these two mechanisms from a mathematical standpoint to illustrate how they achieve adaptation. Possible biomolecular implementations are presented in later chapters.

Integral feedback

In integral feedback systems, a "memory" variable z accounts for the accumulated error between the output of interest $y(t)$, which is affected by an external perturbation u, and its nominal (or desired) steady state value y_0. This accumulated error is then used to change the output y itself through a gain k (Figure 3.9). If the input perturbation u is constant, this feedback loop brings the system output back to the desired value y_0.

To understand why in this system the output $y(t)$, after any constant input perturbation u, tends to y_0 for $t \to \infty$ independently of the (constant) value of u, we write the equations relating the accumulated error z and the output y as obtained from the block diagram of Figure 3.9. The equations representing the system are given by

$$\frac{dz}{dt} = y_0 - y, \qquad y = kz + u,$$

so that the equilibrium is obtained by setting $\dot{z} = 0$, from which we obtain $y = y_0$.

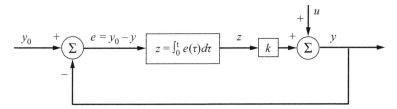

Figure 3.9: A basic block diagram representing a system with integral action. In the diagram, the circles with Σ represent summing junctions, such that the output arrow is a signal given by the sum of the signals associated with the input arrows. The input signals are annotated with a "+" if added or a "−" if subtracted. The desired output y_0 is compared to the actual output y and the resulting error is integrated to yield z. This error is then used to change y. Here, the input u can be viewed as a disturbance input, which perturbs the value of the output y.

That is, the steady state of y does not depend on u. The additional question to answer is whether, after a perturbation u occurs, $y(t)$ tends to y_0 for $t \to \infty$. This is the case if and only if $\dot{z} \to 0$ as $t \to \infty$, which is satisfied if the equilibrium of the system $\dot{z} = -kz - u + y_0$ is asymptotically stable. This, in turn, is satisfied whenever $k > 0$ and u is a constant. Hence, after a constant perturbation u is applied, the system output y approaches its original steady state value y_0, that is, y is robust to constant perturbations.

More generally, a system with integral action can take the form

$$\frac{dx}{dt} = f(x,u), \quad u = (u_1, u_2), \quad y = h(x), \quad \frac{dz}{dt} = y_0 - y, \quad u_2 = k(x,z),$$

in which u_1 is a disturbance input and u_2 is a control input that takes the feedback form $u_2 = k(x,z)$. The steady state value of y, being the solution to $y_0 - y = 0$, does not depend on the disturbance u_1. In turn, y tends to this steady state value for $t \to \infty$ if and only if $\dot{z} \to 0$ as $t \to \infty$. This is the case if z tends to a constant value for $t \to \infty$, which is satisfied if u_1 is a constant and the steady state of the above system is asymptotically stable.

Integral feedback is recognized as a key mechanism of perfect adaptation in biological systems, both at the physiological level and at the cellular level, such as in blood calcium homeostasis [24], in the regulation of tryptophan in *E. coli* [94], in neuronal control of the prefrontal cortex [71], and in *E. coli* chemotaxis [102].

Incoherent feedforward loops

Feedforward motifs (Figure 3.10) are common in transcriptional networks and it has been shown that they are overrepresented in *E. coli* gene transcription networks, compared to other motifs composed of three nodes [4]. Incoherent feedforward circuits represent systems in which the input u directly helps promote the production of the output $y = x_2$ and also acts as a delayed inhibitor of the output

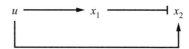

Figure 3.10: Incoherent feedforward loop. The input u affects the output $y = x_2$ through two channels: it indirectly represses it through an intermediate variable x_1 while directly activating it through a different path.

through an intermediate variable x_1. This incoherent counterbalance between positive and negative effects gives rise, under appropriate conditions, to adaptation. A large number of incoherent feedforward loops participate in important biological processes such as the EGF to ERK activation [75], the glucose to insulin release [76], ATP to intracellular calcium release [67], micro-RNA regulation [93], and many others.

Several variants of incoherent feedforward loops exist for perfect adaptation. Here, we consider two main ones, depending on whether the intermediate variable promotes degradation of the output or inhibits its production. An example where the intermediate variable promotes degradation is provided by the "sniffer," which appears in models of neutrophil motion and *Dictyostelium* chemotaxis [101]. In the sniffer, the intermediate variable promotes degradation according to the following differential equation model:

$$\frac{dx_1}{dt} = \alpha u - \gamma x_1, \qquad \frac{dx_2}{dt} = \beta u - \delta x_1 x_2, \qquad y = x_2. \qquad (3.16)$$

In this system, the steady state value of the output x_2 is obtained by setting the time derivatives to zero. Specifically, we have that $\dot{x}_1 = 0$ gives $x_1 = \alpha u / \gamma$ and $\dot{x}_2 = 0$ gives $x_2 = \beta u / (\delta x_1)$. In the case in which $u \neq 0$, these can be combined to yield $x_2 = (\beta \gamma) / (\delta \alpha)$, which is a constant independent of the input u. The linearization of the system at the equilibrium is given by

$$A = \begin{pmatrix} -\gamma & 0 \\ -\delta(\beta\gamma)/(\delta\alpha) & -\delta(\alpha u/\gamma) \end{pmatrix},$$

which has eigenvalues $-\gamma$ and $-\delta(\alpha u/\gamma)$. Since these are both negative, the equilibrium point is asymptotically stable. Note that in the case in which, for example, u goes back to zero after a perturbation, as it is in the case of a pulse, the output x_2 does not necessarily return to its original steady state. That is, this system "adapts" only to constant nonzero input stimuli but is not capable of adapting to pulses. This can be seen from equation (3.16), which admits multiple steady states when $u = 0$. For more details on this "memory" effect, the reader is referred to [91].

A different form for an incoherent feedforward loop is one in which the intermediate variable x_1 inhibits production of the output x_2, such as in the system:

$$\frac{dx_1}{dt} = \alpha u - \gamma x_1, \qquad \frac{dx_2}{dt} = \beta \frac{u}{x_1} - \delta x_2, \qquad y = x_2. \qquad (3.17)$$

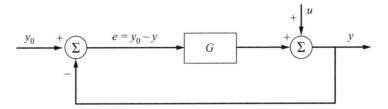

Figure 3.11: High gain feedback. A possible mechanism to attain disturbance attenuation is to feedback the error $y_0 - y$ between the desired output y_0 and the actual output y through a large gain G.

The equilibrium point of this system for a constant nonzero input u is given by setting the time derivatives to zero. From $\dot{x}_1 = 0$, we obtain $x_1 = \alpha u/\gamma$ and from $\dot{x}_2 = 0$ we obtain that $x_2 = \beta u/(\delta x_1)$, which combined together result in $x_2 = (\beta \gamma)/(\delta \alpha)$, which is again a constant independent of the input u.

By calculating the linearization at the equilibrium, one obtains

$$A = \begin{pmatrix} -\gamma & 0 \\ -\gamma^2/(\alpha^2 u) & -\delta \end{pmatrix},$$

whose eigenvalues are given by $-\gamma$ and $-\delta$. Hence, the equilibrium point is asymptotically stable. Further, one can show that the equilibrium point is globally asymptotically stable because the x_1 subsystem is linear, stable, and x_1 approaches a constant value (for constant u) and the x_2 subsystem, in which $\beta u/x_1$ is viewed as an external input is also linear and asymptotically stable.

High gain feedback

Integral feedback and incoherent feedforward loops provide means to obtain exact rejection of constant disturbances. Sometimes, exact rejection is not possible, for example, because the physical constraints of the system do not allow us to implement integral feedback or because the disturbance is not constant with time. In these cases, it may be possible to still attenuate the effect of the disturbance on the output of interest by the use of negative feedback with high gain. To explain this concept, consider the diagram of Figure 3.11.

In a high gain feedback configuration, the error between the output y, perturbed by some exogenous disturbance u, and a desired nominal output y_0 is fed back with a negative sign to produce the output y itself. If $y_0 > y$, this will result in an increase of y, otherwise it will result in a decrease of y. Mathematically, one obtains from the block diagram that

$$y = \frac{u}{1+G} + y_0 \frac{G}{1+G},$$

so that as G increases the (relative) contribution of u on the output of the system can be arbitrarily reduced.

High gain feedback can take a much more general form. Consider a system with $x \in \mathbb{R}^n$ in the form $\dot{x} = f(x)$. We say that this system is *contracting* if any two trajectories starting from different initial conditions exponentially converge to each other as time increases to infinity. A sufficient condition for the system to be contracting is that in some set of coordinates, with matrix transformation denoted Θ, the symmetric part of the linearization matrix (Jacobian)

$$\frac{1}{2}\left(\frac{\partial f}{\partial x} + \frac{\partial f^T}{\partial x}\right)$$

is negative definite. We denote the largest eigenvalue of this matrix by $-\lambda$ for $\lambda > 0$ and call it the contraction rate of the system.

Now, consider the nominal system $\dot{x} = Gf(x)$ for $G > 0$ and its perturbed version $\dot{x}_p = Gf(x_p) + u(t)$. Assume that the input $u(t)$ is bounded everywhere in norm by a constant $C > 0$. If the system is contracting, we have the following robustness result:

$$\|x(t) - x_p(t)\| \leq \chi\|x(0) - x_p(0)\|e^{-G\lambda t} + \frac{\chi C}{\lambda G},$$

in which χ is an upper bound on the condition number of the transformation matrix Θ (ratio between the largest and the smallest eigenvalue of $\Theta^T\Theta$) [62]. Hence, if the perturbed and the nominal systems start from the same initial conditions, the difference between their states can be made arbitrarily small by increasing the gain G. Therefore, the contribution of the disturbance u on the system state can be made arbitrarily small.

A comprehensive treatment of concepts of stability and robustness can be found in standard references [55, 90].

Scale invariance and fold-change detection

Scale invariance is the property by which the output $y(t)$ of the system does not depend on the absolute amplitude of the input $u(t)$ (Figure 3.12). Specifically, consider an adapting system and assume that it preadapted to a constant background input value a, then apply input $a + b$ and let $y(t)$ be the resulting output. Now consider a new background input value pa and let the system preadapt to it. Then apply the input $p(a + b)$ and let $\bar{y}(t)$ be the resulting output. The system has the scale invariance property if $y(t) = \bar{y}(t)$ for all t. This also means that the output responds in the same way to inputs changed by the same multiplicative factor (fold), hence this property is also called *fold-change detection*. Looking at Figure 3.12, the outputs corresponding to the two indicated inputs are identical since the fold change in the input value is equal to b/a in both cases.

Some incoherent feedforward loops can implement the fold-change detection property [35]. As an example, consider the feedforward motif represented by equations (3.17), in which the output is given by $y = x_2$, and consider two inputs:

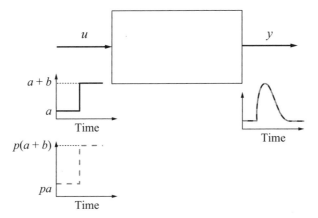

Figure 3.12: Fold-change detection. The output response does not depend on the absolute magnitude of the input but only on the fold change of the input.

$u_1(t) = a$ for $t < t_0$ and $u_1(t) = a + b_1$ for $t \geq t_0$, and $u_2(t) = pa$ for $t < t_0$ and $u_2(t) = pa + pb_1$ for $t \geq t_0$. Assume also that at time t_0 the system is at the steady state, that is, it is preadapted. Hence, we have that the two steady states from which the system starts at $t = t_0$ are given by $x_{1,1} = a\alpha/\gamma$ and $x_{1,2} = pa\alpha/\gamma$ for the x_1 variable and by $x_{2,1} = x_{2,2} = (\beta\gamma)/(\delta\alpha)$ for the x_2 variable. Integrating system (3.17) starting from these initial conditions, we obtain for $t \geq t_0$

$$x_{1,1}(t) = a\frac{\alpha}{\gamma}e^{-\gamma(t-t_0)} + (a+b)(1 - e^{-\gamma(t-t_0)}),$$

$$x_{1,2}(t) = pa\frac{\alpha}{\gamma}e^{-\gamma(t-t_0)} + p(a+b)(1 - e^{-\gamma(t-t_0)}).$$

Using these in the expression of $\dot{x}_2$ in equation (3.17) gives the differential equations that $x_{2,1}(t)$ and $x_{2,2}(t)$ obey for $t \geq t_0$ as

$$\frac{dx_{2,1}}{dt} = \frac{\beta(a+b)}{a\frac{\alpha}{\gamma}e^{-\gamma(t-t_0)} + (a+b)(1 - e^{-\gamma(t-t_0)})} - \delta x_{2,1}, \qquad x_{2,1}(t_0) = (\beta\gamma)/(\delta\alpha)$$

and

$$\frac{dx_{2,2}}{dt} = \frac{p\beta(a+b)}{pa\frac{\alpha}{\gamma}e^{-\gamma(t-t_0)} + p(a+b)(1 - e^{-\gamma(t-t_0)})} - \delta x_{2,2}, \qquad x_{2,2}(t_0) = (\beta\gamma)/(\delta\alpha),$$

which gives $x_{2,1}(t) = x_{2,2}(t)$ for all $t \geq t_0$. Hence, the system responds exactly the same way after changes in the input of the same fold. The output response is not dependent on the scale of the input but only on its shape.

3.3 Oscillatory behavior

In addition to equilibrium behavior, a variety of cellular processes involve oscillatory behavior in which the system state is constantly changing, but in a repeating pattern. Two examples of biological oscillations are the cell cycle and circadian rhythm. Both of these dynamic behaviors involve repeating changes in the concentrations of various proteins, complexes and other molecular species in the cell, though they are very different in their operation. In this section we discuss some of the underlying ideas for how to model this type of oscillatory behavior, focusing on those types of oscillations that are most common in biomolecular systems.

Biomolecular oscillators

Biological systems have a number of natural oscillatory processes that govern the behavior of subsystems and whole organisms. These range from internal oscillations within cells to the oscillatory nature of the beating heart to various tremors and other undesirable oscillations in the neuro-muscular system. At the biomolecular level, two of the most studied classes of oscillations are the cell cycle and circadian rhythm.

The cell cycle consists of a set of "phases" that govern the duplication and division of cells into two new cells:

- G1 phase - gap phase, terminated by "G1 checkpoint";
- S phase - synthesis phase (DNA replication);
- G2 phase - gap phase, terminated by "G2 checkpoint";
- M - mitosis (cell division).

The cell goes through these stages in a cyclical fashion, with the different enzymes and pathways active in different phases. The cell cycle is regulated by many different proteins, often divided into two major classes. *Cyclins* are a class of proteins that sense environmental conditions internal and external to the cell and are also used to implement various logical operations that control transition out of the G1 and G2 phases. *Cyclin dependent kinases* (CDKs)are proteins that serve as "actuators" by turning on various pathways during different cell cycles.

An example of the control circuitry of the cell cycle for the bacterium *Caulobacter crescentus* (henceforth *Caulobacter*) is shown in Figure 3.13 [59]. This organism uses a variety of different biomolecular mechanisms, including transcriptional activation and repression, positive autoregulation (CtrA), phosphotransfer and methylation of DNA. The cell cycle is an example of an oscillator that does not have a fixed period. Instead, the length of the individual phases and the transitioning of the different phases are determined by the environmental conditions. As one example, the cell division time for *E. coli* can vary between 20 and 90 minutes due to changes in nutrient concentrations, temperature or other external factors.

A different type of oscillation is the highly regular pattern encoding in circadian rhythm, which repeats with a period of roughly 24 hours. The observation of circadian rhythms dates as far back as 400 BCE, when Androsthenes described observations of daily leaf movements of the tamarind tree [69]. There are three

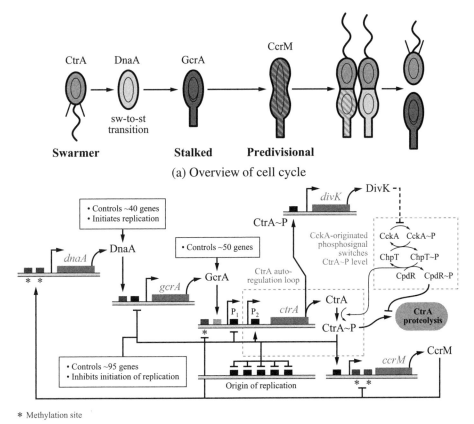

(a) Overview of cell cycle

(b) Molecular mechanisms controlling cell cycle

Figure 3.13: The *Caulobacter crescentus* cell cycle. (a) *Caulobacter* cells divide asymmetrically into a stalked cell, which is attached to a surface, and a swarmer cell that is motile. The swarmer cells can become stalked cells in a new location and begin the cell cycle anew. The transcriptional regulators CtrA, DnaA and GcrA are the primary factors that control the various phases of the cell cycle. (b) The genetic circuitry controlling the cell cycle consists of a large variety of regulatory mechanisms, including transcriptional regulation and post-translational regulation. Figure adapted from [59].

defining characteristics associated with circadian rhythm: (1) the time to complete one cycle is approximately 24 hours, (2) the rhythm is endogenously generated and self-sustaining, and (3) the period remains relatively constant under changes in ambient temperature. Oscillations that have these properties appear in many different organisms, including microorganisms, plants, insects and mammals. Some common features of the circuitry implementing circadian rhythms in these organisms is the combination of positive and negative feedback loops, often with the positive elements activating the expression of clock genes and the negative elements repressing the positive elements [11]. Figure 3.14 shows some of the different organisms in which circadian oscillations can be found and the primary genes

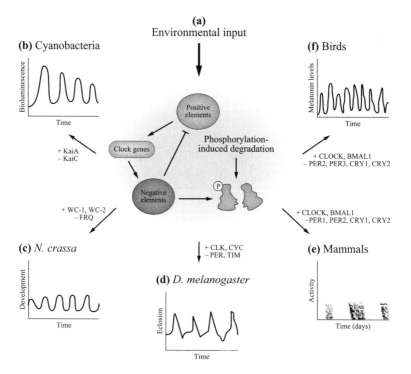

Figure 3.14: Overview of mechanisms for circadian rhythm in different organisms. Circadian rhythms are found in many different classes of organisms. A common pattern is a combination of positive and negative feedback, as shown in the center of the figure. Driven by environmental inputs (a), a variety of different genes are used to implement these positive and negative elements (b–f). Figure adapted from [11].

responsible for different positive and negative factors.

Clocks, oscillators and limit cycles

To begin our study of oscillators, we consider a nonlinear model of the system described by the differential equation

$$\frac{dx}{dt} = f(x,\theta,u), \qquad y = h(x,\theta),$$

where $x \in \mathbb{R}^n$ represents the state of the system, $u \in \mathbb{R}^q$ represents the external inputs, $y \in \mathbb{R}^m$ represents the (measured) outputs and $\theta \in \mathbb{R}^p$ represents the model parameters. We say that a solution $(x(t), u(t))$ is *oscillatory with period T* if $y(t + T) = y(t)$. For simplicity, we will often assume that $p = q = 1$, so that we have a single input and single output, but most of the results can be generalized to the multi-input, multi-output case.

There are multiple ways in which a solution can be oscillatory. One of the simplest is that the input $u(t)$ is oscillatory, in which case we say that we have a *forced oscillation*. In the case of a stable linear system with one input and one output, an input of the form $u(t) = A \sin \omega t$ will lead, after the transient due to initial conditions has died out, to an output of the form $y(t) = M \cdot A \sin(\omega t + \phi)$ where M and ϕ represent the gain and phase of the system (at frequency ω). In the case of a nonlinear system, if the output is periodic with the same period then we can write it in terms of a set of harmonics,

$$y(t) = B_0 + B_1 \sin(\omega t + \phi_1) + B_2 \sin(2\omega t + \phi_2) + \cdots .$$

The term B_0 represents the average value of the output (also called the bias), the terms B_i are the magnitudes of the ith harmonic and ϕ_i are the phases of the harmonics (relative to the input). The *oscillation frequency* ω is given by $\omega = 2\pi/T$ where T is the oscillation period.

A different situation occurs when we have no input (or a constant input) and still obtain an oscillatory output. In this case we say that the system has a *self-sustained oscillation*. This type of behavior is what is required for oscillations such as the cell cycle and circadian rhythm, where there is either no obvious forcing function or the forcing function is removed but the oscillation persists. If we assume that the input is constant, $u(t) = A_0$, then we are particularly interested in how the period T (or equivalently the frequency ω), amplitudes B_i and phases ϕ_i depend on the input A_0 and system parameters θ.

To simplify our notation slightly, we consider a system of the form

$$\frac{dx}{dt} = f(x,\theta), \qquad y = h(x,\theta), \tag{3.18}$$

where the input is ignored (or taken to be one of the constant parameters) in the analysis that follows. We have focused on the oscillatory nature of the output $y(t)$ thus far, but we note that if the states $x(t)$ are periodic then the output is as well, and this is the most common case. Hence we will often talk about the *system* being oscillatory, by which we mean that there is a solution for the dynamics in which the state satisfies $x(t + T) = x(t)$.

More formally, we say that a closed curve $\Gamma \in \mathbb{R}^n$ is an *orbit* if trajectories that start on Γ remain on Γ for all time and if Γ is not an equilibrium point of the system. As in the case of equilibrium points, we say that the orbit is *stable* if trajectories that start near Γ stay near Γ, *asymptotically stable* if in addition nearby trajectories approach Γ as $t \to \infty$ and *unstable* if it is not stable. The orbit Γ is periodic with period T if for any $x(t) \in \Gamma$, $x(t + T) = x(t)$.

There are many different types of periodic orbits that can occur in a system whose dynamics are modeled as in equation (3.18). A *harmonic oscillator* references to a system that oscillates around an equilibrium point, but does not (usually) get near the equilibrium point. The classical harmonic oscillator is a linear system of the form

$$\frac{d}{dt}\begin{pmatrix} x_1 \\ x_2 \end{pmatrix} = \begin{pmatrix} 0 & \omega \\ -\omega & 0 \end{pmatrix}\begin{pmatrix} x_1 \\ x_2 \end{pmatrix},$$

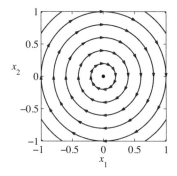

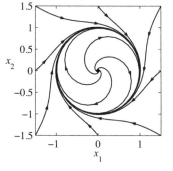

(a) Linear harmonic oscillator   (b) Nonlinear harmonic oscillator

Figure 3.15: Examples of harmonic oscillators.

whose solutions are given by

$$\begin{pmatrix} x_1(t) \\ x_2(t) \end{pmatrix} = \begin{pmatrix} \cos \omega t & \sin \omega t \\ -\sin \omega t & \cos \omega t \end{pmatrix} \begin{pmatrix} x_1(0) \\ x_2(0) \end{pmatrix}.$$

The frequency of this oscillation is fixed, but the amplitude depends on the values of the initial conditions, as shown in Figure 3.15a. Note that this system has a single equilibrium point at $x = (0,0)$ and the eigenvalues of the equilibrium point have zero real part, so trajectories neither expand nor contract, but simply oscillate.

An example of a nonlinear harmonic oscillator is given by the equation

$$\frac{dx_1}{dt} = x_2 + x_1(1 - x_1^2 - x_2^2), \qquad \frac{dx_2}{dt} = -x_1 + x_2(1 - x_1^2 - x_2^2). \qquad (3.19)$$

This system has an equilibrium point at $x = (0,0)$, but the linearization of this equilibrium point is unstable. The phase portrait in Figure 3.15b shows that the solutions in the phase plane converge to a circular trajectory. In the time domain this corresponds to an oscillatory solution. Mathematically the circle is called a *limit cycle*. Note that in this case, the solution for any initial condition approaches the limit cycle and the amplitude and frequency of oscillation "in steady state" (once we have reached the limit cycle) are independent of the initial condition.

A different type of oscillation can occur in nonlinear systems in which the equilibrium points are saddle points, having both stable and unstable eigenvalues. Of particular interest is the case where the stable and unstable orbits of one or more equilibrium points join together. Two such situations are shown in Figure 3.16. The figure on the left is an example of a *homoclinic orbit*. In this system, trajectories that start near the equilibrium point quickly diverge away (in the directions corresponding to the unstable eigenvalues) and then slowly return to the equilibrium point along the stable directions. If the initial conditions are chosen to be precisely on the homoclinic orbit Γ then the system slowly converges to the equilibrium

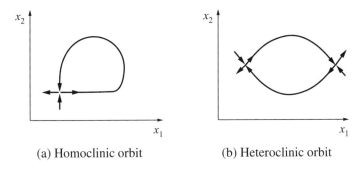

(a) Homoclinic orbit   (b) Heteroclinic orbit

Figure 3.16: Homoclinic and heteroclinic orbits.

point, but in practice there are often disturbances present that will perturb the system off of the orbit and trigger a "burst" in which the system rapidly escapes from the equilibrium point and then slowly converges again. A somewhat similar type of orbit is a *heteroclinic orbit*, in which the orbit connects two different equilibrium points, as shown in Figure 3.16b.

An example of a system with a homoclinic orbit is given by the system

$$\frac{dx_1}{dt} = x_2, \qquad \frac{dx_2}{dt} = x_1 - x_1^3. \tag{3.20}$$

The phase portrait and time domain solutions are shown in Figure 3.17. In this system, there are periodic orbits both inside and outside the two homoclinic cycles (left and right). Note that the trajectory we have chosen to plot in the time domain has the property that it rapidly moves away from the equilibrium point and then slowly reconverges to the equilibrium point, before being carried away again. This type of oscillation, in which one slowly returns to an equilibrium point before rapidly diverging is often called a *relaxation oscillation*. Note that for this system, there are also oscillations that look more like the harmonic oscillator case described above, in which we oscillate around the unstable equilibrium points at $x = (\pm 1, 0)$.

Example 3.8 (Glycolytic oscillations). Glycolysis is one of the principal metabolic networks involved in energy production. It is a sequence of enzyme-catalyzed reactions that converts sugar into pyruvate, which is then further degraded to alcohol (in yeast fermentation) and lactic acid (in muscles) under anaerobic conditions, and ATP (the cell's major energy supply) is produced as a result. Both damped and sustained oscillations have been observed. Damped oscillations were first reported by [23] while sustained oscillations in yeast cell free extracts were observed in [42, 81].

Here we introduce the basic motif that is known to be at the core of this oscillatory phenomenon. Specifically, a substrate S is converted to a product P, which, in turn, acts as an enzyme catalyzing the conversion of S into P. This is an example of autocatalysis, in which a product is required for its own production. A simple

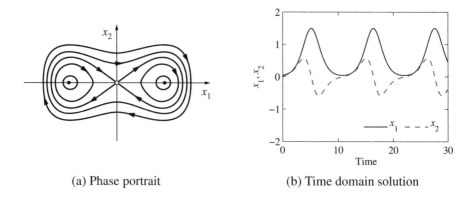

(a) Phase portrait   (b) Time domain solution

Figure 3.17: Example of a homoclinic orbit.

differential equation model of this system can be written as

$$\frac{dS}{dt} = v_0 - v_1, \qquad \frac{dP}{dt} = v_1 - v_2, \tag{3.21}$$

in which v_0 is a constant production rate and

$$v_1 = S\,F(P), \quad F(P) = \frac{\alpha(P/K)^2}{1+(P/K)^2}, \qquad v_2 = k_2 P,$$

where $F(P)$ is the standard Hill function. Under the assumption that $K \gg P$, we have $F(P) \approx k_1 P^2$, in which we have defined $k_1 := \alpha/K^2$. This second-order system admits a stable limit cycle under suitable parameter conditions (Figure 3.18).   ∇

One central question when analyzing the dynamical model of a given system is to establish whether the model constructed admits sustained oscillations. This

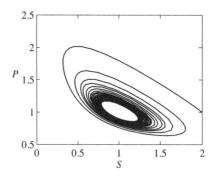

Figure 3.18: Oscillations in the glycolysis system. Parameters are $v_0 = 1$, $k_1 = 1$, and $k_2 = 1.00001$.

way we can validate or disprove models of biomolecular systems that are known to exhibit sustained oscillations. At the same time, we can provide design guidelines for engineering biological circuits that function as clocks, as we will see in Chapter 5. With this respect, it is particularly important to determine parameter conditions that are required and/or sufficient to obtain periodic behavior. To analyze these sorts of questions, we need to introduce tools that allow us to infer the existence and robustness of a limit cycle from a differential equation model.

In order to proceed, we first introduce the concept of ω-limit set of a point p, denoted $\omega(p)$. Basically, the ω-limit set $\omega(p)$ represents the set of all points to which the trajectory of the system starting from p tends as time approaches infinity. This is formally defined in the following definition.

Definition 3.1. A point $\bar{x} \in \mathbb{R}^n$ is called an *ω-limit point* of $p \in \mathbb{R}^n$ if there is a sequence of times $\{t_i\}$ with $t_i \to \infty$ for $i \to \infty$ such that $x(t_i, p) \to \bar{x}$ as $i \to \infty$. The *ω-limit set* of p, denoted $\omega(p)$, is the set of all ω-limit points of p.

The ω-limit set of a system has several relevant properties, among which are the facts that it cannot be empty and that it must be a connected set.

Limit cycles in the plane

Before studying periodic behavior of systems in $\mathbb{R}^n$, we study the behavior of systems in $\mathbb{R}^2$. Several high-dimensional systems can often be well approximated by systems in two dimensions by, for example, employing quasi-steady state approximations. For systems in $\mathbb{R}^2$, we will see that there are easy-to-check conditions that guarantee the existence of a limit cycle.

The first result provides a simple check to rule out periodic solutions for systems in $\mathbb{R}^2$. Specifically, let $x \in \mathbb{R}^2$ and consider

$$\frac{dx_1}{dt} = f_1(x_1, x_2), \qquad \frac{dx_2}{dt} = f_2(x_1, x_2), \qquad (3.22)$$

in which the functions $f_i : \mathbb{R}^2 \to \mathbb{R}^2$ for $i = 1, 2$ are smooth. Then, we have the following:

Theorem 3.2 (Bendixson's criterion). *Let D be a simply connected region in $\mathbb{R}^2$ (i.e., there are no holes in D). If the expression*

$$\frac{\partial f_1}{\partial x_1} + \frac{\partial f_2}{\partial x_2}$$

is not identically zero and does not change sign in D, then system (3.22) has no closed orbits that lie entirely in D.

Example 3.9. Consider the system

$$\frac{dx_1}{dt} = -x_2^3 + \delta x_1^3, \qquad \frac{dx_2}{dt} = x_1^3,$$

with $\delta \geq 0$. We can compute

$$\frac{\partial f_1}{\partial x_1} + \frac{\partial f_2}{\partial x_2} = 3\delta x_1^2,$$

which is not identically zero and does not change sign over all of $\mathbb{R}^2$ when $\delta \neq 0$. If $\delta \neq 0$, we can thus conclude from Bendixson's criterion that there are no periodic solutions. We leave it as an exercise to investigate what happens when $\delta = 0$ (Exercise 3.5). ∇

The following theorem completely characterizes the ω-limit set of any point for a system in $\mathbb{R}^2$.

Theorem 3.3 (Poincaré-Bendixson). *Let M be a bounded and closed positively invariant region for the system $\dot{x} = f(x)$ with $x(0) \in M$ (i.e., any trajectory that starts in M stays in M for all $t \geq 0$). Assume that there are finitely many equilibrium points in M. Let $p \in M$, then one of the following possibilities holds for $\omega(p)$:*

(i) $\omega(p)$ is an equilibrium point;

(ii) $\omega(p)$ is a closed orbit;

(iii) $\omega(p)$ consists of a finite number of equilibrium points and orbits, each start-ing (for $t = 0$) and ending (for $t \to \infty$) at one of the fixed points.

This theorem has two important consequences:

1. If the system does not have equilibrium points in M, since $\omega(p)$ is not empty, it must be a periodic solution;

2. If there is only one equilibrium point in M and it is unstable and not a saddle (i.e., the eigenvalues of the linearization at the equilibrium point are both positive), then $\omega(p)$ is a periodic solution.

We will employ this result in Chapter 5 to determine parameter conditions under which activator-repressor circuits admit sustained oscillations.

Limit cycles in $\mathbb{R}^n$

The results above hold only for systems in two dimensions. However, there have been extensions of the Poincaré-Bendixson theorem to systems with special struc-ture in $\mathbb{R}^n$. In particular, we have the following result, which can be stated as fol-lows under some mild technical assumptions, which we omit here.

Theorem 3.4 (Hastings et al. [40]). *Consider a system $\dot{x} = f(x)$, which is of the form*

$$\dot{x}_1 = f_1(x_n, x_1),$$
$$\dot{x}_j = f_j(x_{j-1}, x_j), \ 2 \leq j \leq n$$

on the set M defined by $x_i \geq 0$ for all i with the following inequalities holding in M:

(i) $\frac{\partial f_i}{\partial x_i} < 0$ and $\frac{\partial f_i}{\partial x_{i-1}} > 0$, for $2 \leq i \leq n$, and $\frac{\partial f_1}{\partial x_n} < 0$;

(ii) $f_i(0,0) \geq 0$ and $f_1(x_n,0) > 0$ for all $x_n \geq 0$;

(iii) The system has a unique equilibrium point $x^* = (x_1^*, ..., x_n^*)$ in M such that $f_1(x_n, x_1) < 0$ if $x_n > x_n^*$ and $x_1 > x_1^*$, while $f_1(x_n, x_1) > 0$ if $x_n < x_n^*$ and $x_1 < x_1^*$;

(iv) $\frac{\partial f_1}{\partial x_1}$ is bounded above in M.

Then, if the Jacobian of f at x^* has no repeated eigenvalues and has any eigenvalue with positive real part, then the system has a non-constant periodic solution in M.

This theorem states that for a system with cyclic structure in which the cycle "has negative loop gain," the instability of the equilibrium point (under some technical assumption) is equivalent to the existence of a periodic solution. This theorem, however, does not provide information about whether the orbit is attractive or not, that is, whether it is an ω-limit set of any point in M. This stability result is implied by a general theorem, which can be stated as follows under some mild technical assumptions, which we omit here.

Theorem 3.5 (Mallet-Paret and Smith [64]). *Consider the system $\dot{x} = f(x)$ with the following cyclic feedback structure*

$$\dot{x}_1 = f_1(x_n, x_1),$$
$$\dot{x}_j = f_j(x_{j-1}, x_j), \quad 2 \leq j \leq n$$

on a set M defined by $x_i \geq 0$ for all i with all trajectories starting in M bounded for $t \geq 0$. Then, the ω-limit set $\omega(p)$ of any point $p \in M$ can be one of the following:

(i) An equilibrium point;

(ii) A non-constant periodic orbit;

(iii) A set of equilibrium points connected by homoclinic or heteroclinic orbits.

As a consequence of the theorem, we have that for a system with cyclic feedback structure that admits one equilibrium point only and at which the linearization has all eigenvalues with positive real part, the ω-limit set must be a periodic orbit.

In Chapter 5, we will apply these results to determine parameter conditions that make loop circuits with state in $\mathbb{R}^n$ admit a limit cycle.

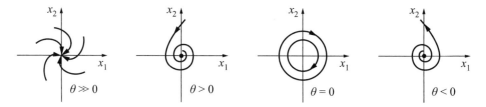

Figure 3.19: Phase portraits for a linear system as parameter θ changes. When θ is positive and large in absolute value, the eigenvalues are real and negative, and the response is not oscillatory (overdamped). When θ is positive but not too large in absolute value, the eigenvalues are complex with negative real part and damped oscillations arise (underdamped). When $\theta = 0$, the system displays oscillatory solutions, while when $\theta < 0$, the equilibrium point becomes unstable and trajectories diverge.

3.4 Bifurcations

Another important property of nonlinear systems is how their behavior changes as the parameters governing the dynamics change. We can study this in the context of models by exploring how the location of equilibrium points, their stability, their regions of attraction and other dynamic phenomena, such as limit cycles, vary based on the values of the parameters in the model.

Parametric stability

Consider a differential equation of the form

$$\frac{dx}{dt} = f(x, \theta), \quad x \in \mathbb{R}^n, \theta \in \mathbb{R}^p, \tag{3.23}$$

where x is the state and θ is a set of parameters that describe the family of equations. The equilibrium solutions satisfy

$$f(x, \theta) = 0,$$

and as θ is varied, the corresponding solutions $x_e(\theta)$ can also vary. We say that the system (3.23) has a *bifurcation* at $\theta = \theta^*$ if the behavior of the system changes qualitatively at θ^*. This can occur either because of a change in stability type or because of a change in the number of solutions at a given value of θ.

As an example of a bifurcation, consider the linear system

$$\frac{dx_1}{dt} = x_2, \qquad \frac{dx_2}{dt} = -kx_1 - \theta x_2,$$

where $k > 0$ is fixed and θ is our bifurcation parameter. Figure 3.19 shows the phase portraits for different values of θ. We see that at $\theta = 0$ the system transitions

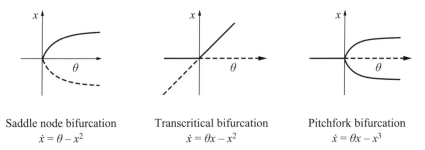

Saddle node bifurcation   Transcritical bifurcation   Pitchfork bifurcation
$\dot{x} = \theta - x^2$ $\dot{x} = \theta x - x^2$ $\dot{x} = \theta x - x^3$

Figure 3.20: Bifurcation diagrams for some common bifurcations. In a *saddle node bifurcation*, as θ decreases a stable and an unstable equilibrium point approach each other and then "collide" for $\theta = 0$ and annihilate each other. In a *transcritical bifurcation*, a stable and an unstable equilibrum point approach each other, and then intersect at $\theta = 0$, swapping their stability. In a *pitchfork bifurcation*, a unique stable equilibrium point for $\theta < 0$ gives rise to three equilibria at the point $\theta = 0$, of which two are stable and one is unstable.

from a single stable equilibrium point at the origin to having an unstable equilibrium. Hence, as θ goes from negative to positive values, the behavior of the system changes in a significant way, indicating a bifurcation.

A common way to visualize a bifurcation is through the use of a *bifurcation diagram*. To create a bifurcation diagram, we choose a function $y = h(x)$ such that the value of y at an equilibrium point has some useful meaning for the question we are studying. We then plot the value of $y_e = h(x_e(\theta))$ as a function of θ for all equilibria that exist for a given parameter value θ. By convention, we use dashed lines if the corresponding equilibrium point is unstable and solid lines otherwise. Figure 3.20 shows examples of some common bifurcation diagrams. Note that for some types of bifurcations, such as the pitchfork bifurcation, there exist values of θ where there is more than one equilibrium point. A system that exhibits this type of behavior is said to be *multistable*. A common case is when there are two stable equilibria, in which case the system is said to be *bistable*. We will see an example of this in Chapter 5.

Hopf bifurcation

The bifurcations discussed above involved bifurcation of equilibrium points. Another type of bifurcation that can occur is when a system with an equilibrium point admits a limit cycle as a parameter is changed through a critical value. The Hopf bifurcation theorem provides a technique that is often used to understand whether a system admits a periodic orbit when some parameter is varied. Usually, such an orbit is a small amplitude periodic orbit that is present in the close vicinity of an unstable equilibrium point.

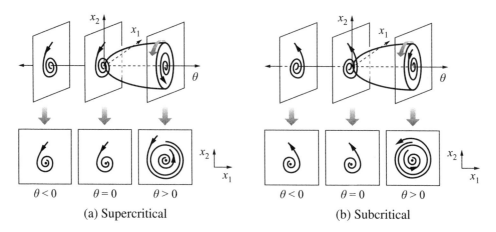

(a) Supercritical   (b) Subcritical

Figure 3.21: Hopf bifurcation. (a) As θ increases a stable limit cycle appears. (b) As θ increases a periodic orbit appears but it is unstable. Figure adapted from [100].

Consider the system dependent on a parameter α:

$$\frac{dx}{dt} = g(x, \alpha), \qquad x \in \mathbb{R}^n, \qquad \alpha \in \mathbb{R},$$

and assume that at the equilibrium point x_e corresponding to $\alpha = \alpha_0$ (i.e., $g(x_e, \alpha_0) = 0$), the linearization $\partial g/\partial x$ evaluated at (x_e, α_0) has a pair of (nonzero) imaginary eigenvalues with the remaining eigenvalues having negative real parts. Define the new parameter $\theta := \alpha - \alpha_0$ and redefine the system as

$$\frac{dx}{dt} = f(x, \theta) =: g(x, \theta + \alpha_0),$$

so that the linearization $\partial f/\partial x$ evaluated at $(x_e, 0)$ has a pair of (nonzero) imaginary eigenvalues with the remaining eigenvalues having negative real parts. Denote by $\lambda(\theta) = \beta(\theta) + i\omega(\theta)$ the eigenvalue such that $\beta(0) = 0$. Then, if $\partial \beta/\partial \theta$ evaluated at $\theta = 0$ is not zero, the system admits a small amplitude almost sinusoidal periodic orbit for θ small enough and the system is said to go through a Hopf bifurcation at $\theta = 0$. If the small amplitude periodic orbit is stable, the Hopf bifurcation is said to be *supercritical*, while if it is unstable it is said to be *subcritical*. Figure 3.21 shows diagrams corresponding to these bifurcations.

In order to determine whether a Hopf bifurcation is supercritical or subcritical, it is necessary to calculate a "curvature" coefficient, for which there are formulas (Marsden and McCracken [66]) and available bifurcation software, such as AUTO. In practice, it is often enough to calculate the value $\bar{\alpha}$ of the parameter at which the Hopf bifurcation occurs and simulate the system for values of the parameter α close to $\bar{\alpha}$. If a small amplitude limit cycle appears, then the bifurcation is most likely supercritical.

Example 3.10 (Glycolytic oscillations). Recalling the model of glycolytic oscillations given in (3.21), we ask whether such an oscillator goes through a Hopf bifurcation. In order to answer this question, we consider again the expression of the eigenvalues

$$\lambda_{1,2} = \frac{\text{tr}(J) \pm \sqrt{\text{tr}(J)^2 - 4\det(J)}}{2},$$

in which

$$\text{tr}(J) = k_2 - k_1 \left(\frac{v_0}{k_2}\right)^2 \quad \text{and} \quad \det(J) = k_1 \left(\frac{v_0}{k_2}\right)^2.$$

The eigenvalues are imaginary if $\text{tr}(J) = 0$, that is, if $k_1 = k_2^3/v_0^2$. Furthermore, the frequency of oscillations is given by $\omega = \sqrt{4\det(J)} = \sqrt{4k_1(v_0/k_2)^2}$. Therefore, this system goes through a Hopf bifurcation as the parameter k_1 approaches k_2^3/v_0^2. When $k_1 \approx k_2^3/v_0^2$, an approximately sinusoidal oscillation appears. When k_1 is large, the Hopf bifurcation theorem does not imply the existence of a periodic solution. This is because the Hopf theorem provides only local results.   ▽

The Hopf bifurcation theorem is based on center manifold theory for nonlinear dynamical systems. For a rigorous treatment of Hopf bifurcation it is thus necessary to study center manifold theory first, which is outside the scope of this text. For details, the reader is referred to standard text in dynamical systems [100, 39].

In Chapter 5, we will illustrate how to employ Hopf bifurcation to understand one of the key design principles of clocks based on two interacting species, an activator and a repressor.

3.5 Model reduction techniques

The techniques that we have developed in this chapter can be applied to a wide variety of dynamical systems. However, many of the methods require significant computation and hence we would like to reduce the complexity of the models as much as possible before applying them. In this section, we review methods for doing such a reduction in the complexity of models. Most of the techniques are based on the common idea that if we are interested in the slower time scale dynamics of a system, the fast time scale dynamics can be approximated by their equilibrium solutions. This idea was introduced in Chapter 2 in the context of reduced order mechanisms; we present a more detailed mathematical analysis of such systems here.

The mathematical analysis of systems with multiple time scales and the consequent model order reduction is called *singular perturbation theory*. In particular, we are concerned with systems that have processes evolving on both fast and slow time scales and that can be written in a standard form, which we now introduce.

Let $(x, y) \in D := D_x \times D_y \subset \mathbb{R}^n \times \mathbb{R}^m$ and consider the vector field

$$
\begin{aligned}
\frac{dx}{dt} &= f(x, y, \epsilon), && x(0) = x_0, \\
\epsilon \frac{dy}{dt} &= g(x, y, \epsilon), && y(0) = y_0,
\end{aligned}
\tag{3.24}
$$

in which $0 < \epsilon \ll 1$ is a small parameter and both $f(x, y, 0)$ and $g(x, y, 0)$ are well-defined. Since $\epsilon \ll 1$, the rate of change of y can be much larger than the rate of change of x, resulting in y dynamics that are much faster than the x dynamics. That is, this system has a slow time scale evolution (in x) and a fast time scale evolution (in y), so that x is called the slow variable and y is called the fast variable.

If we are interested only in the slower time scale, then the above system can be approximated (under suitable conditions) by the *reduced system*

$$
\begin{aligned}
\frac{d\bar{x}}{dt} &= f(\bar{x}, \bar{y}, 0), && \bar{x}(0) = x_0, \\
0 &= g(\bar{x}, \bar{y}, 0),
\end{aligned}
$$

in which we have set $\epsilon = 0$. Let $y = h(x)$ denote the locally unique solution of $g(x, y, 0) = 0$. The manifold of (x, y) pairs where $y = h(x)$ is called the *slow manifold*. The *implicit function theorem* [65] shows that this solution exists whenever $\partial g / \partial y$ is, at least locally, nonsingular. In fact, in such a case we have

$$
\frac{dh}{dx} = -\frac{\partial g}{\partial y}^{-1} \frac{\partial g}{\partial x}.
$$

We can rewrite the dynamics of x in the reduced system as

$$
\frac{d\bar{x}}{dt} = f(\bar{x}, h(\bar{x}), 0), \qquad \bar{x}(0) = x_0.
$$

We seek to determine under what conditions the solution $x(t)$ is "close" to the solution $\bar{x}(t)$ of the reduced system. This problem can be addressed by analyzing the fast dynamics, that is, the dynamics of the system in the fast time scale $\tau = t / \epsilon$. In this case, we have that

$$
\frac{dx}{d\tau} = \epsilon f(x, y, \epsilon), \qquad \frac{dy}{d\tau} = g(x, y, \epsilon), \qquad (x(0), y(0)) = (x_0, y_0),
$$

so that when $\epsilon \ll 1$, $x(\tau)$ does not appreciably change. Therefore, the above system in the τ time scale can be well approximated by the system

$$
\frac{dy}{d\tau} = g(x_0, y, 0), \qquad y(0) = y_0,
$$

in which x is "frozen" at the initial condition x_0. This system is usually referred to as the *boundary layer* system. For this system, the point $y = h(x_0)$ is an equilibrium point. Such an equilibrium point is asymptotically stable if $y(\tau)$ converges to $h(x_0)$ as $\tau \to \infty$. In this case, the solution $(x(t), y(t))$ of the original system approaches $(\bar{x}(t), h(\bar{x}(t)))$. This qualitative explanation is more precisely captured by the following singular perturbation theorem under some mild technical assumptions, which we omit here [55].

Theorem 3.6. *Assume that*

$$Real\left(\lambda\left(\frac{\partial}{\partial y}g(x,y)\Big|_{y=h(x)}\right)\right) < 0$$

uniformly for $x \in D_x$. Let the solution of the reduced system be uniquely defined for $t \in [0, t_f]$. Then, for all $t_b \in (0, t_f]$ there are constants $\epsilon^* > 0$ and $M > 0$, and a set $\Omega \subseteq D$ such that

$$\|x(t) - \bar{x}(t)\| \leq M\epsilon \text{ for } t \in [0, t_f],$$
$$\|y(t) - h(\bar{x}(t))\| \leq M\epsilon \text{ for } t \in [t_b, t_f],$$

provided $\epsilon < \epsilon^*$ and $(x_0, y_0) \in \Omega$.

Example 3.11 (Hill function). In Section 2.1, we obtained the expression of the Hill function by making a quasi-steady state approximation on the dynamics of reversible binding reactions. Here, we illustrate how Hill function expressions can be derived by a formal application of singular perturbation theory. Specifically, consider the simple binding scenario of a transcription factor X with DNA promoter sites p. Assume that such a transcription factor is acting as an activator of the promoter and let Y be the protein expressed under promoter p. Assume further that X dimerizes before binding to promoter p. The reaction equations describing this system are given by

$$X + X \underset{k_2}{\overset{k_1}{\rightleftharpoons}} X_2, \qquad X_2 + p \underset{d}{\overset{a}{\rightleftharpoons}} C, \qquad C \xrightarrow{k_f} m_Y + C,$$

$$m_Y \xrightarrow{\kappa} m_Y + Y, \qquad m_Y \xrightarrow{\delta} \emptyset, \qquad Y \xrightarrow{\gamma} \emptyset, \qquad p + C = p_{tot}.$$

The corresponding differential equation model is given by

$$\frac{dX_2}{dt} = k_1 X^2 - k_2 X_2 - a X_2(p_{tot} - C) + dC, \qquad \frac{dm_Y}{dt} = k_f C - \delta m_Y,$$
$$\frac{dC}{dt} = a X_2(p_{tot} - C) - dC, \qquad \frac{dY}{dt} = \kappa m_Y - \gamma Y,$$

in which we view $X(t)$ as an input to the system. We will see later in Chapter 6 that the dynamics of the input $X(t)$ will be "perturbed" by the physical process of reversible binding that makes it possible for the system to take X as an input.

Since all the binding reactions are much faster than mRNA and protein production and decay, we have that $k_2, d \gg k_f, \kappa, \delta, \gamma$. Let $K_m := k_2/k_1$, $K_d := d/a$, $c := k_2/d$, and $\epsilon := \gamma/d$. Then, we can rewrite the above system by using the substitutions

$$d = \frac{\gamma}{\epsilon}, \qquad a = \frac{\gamma}{K_d \epsilon}, \qquad k_1 = c\frac{\gamma}{K_m \epsilon}, \qquad k_2 = c\frac{\gamma}{\epsilon},$$

so that we obtain

$$\epsilon\frac{dX_2}{dt} = c\frac{\gamma}{K_m}X^2 - c\gamma X_2 - \frac{\gamma}{K_d}X_2(p_{tot} - C) + \gamma C, \qquad \frac{dm_Y}{dt} = k_f C - \delta m_Y,$$

$$\epsilon\frac{dC}{dt} = \frac{\gamma}{K_d}X_2(p_{tot} - C) - \gamma C, \qquad\qquad \frac{dY}{dt} = \kappa m_Y - \gamma Y.$$

This system is in the standard singular perturbation form (3.24). As an exercise, the reader can verify that the slow manifold is locally asymptotically stable (see Exercise 3.10). The slow manifold is obtained by setting $\epsilon = 0$ and determines X_2 and C as functions of X. These functions are given by

$$X_2 = \frac{X^2}{K_m}, \qquad C = \frac{p_{tot}X^2/(K_m K_d)}{1 + X^2/(K_m K_d)}.$$

As a consequence, the reduced system becomes

$$\frac{dm_Y}{dt} = k_f\frac{p_{tot}X^2/(K_m K_d)}{1 + X^2/(K_m K_d)} - \delta m_Y,$$

$$\frac{dY}{dt} = \kappa m_Y - \gamma Y,$$

which is the familiar expression for the dynamics of gene expression with an activator as derived in Section 2.1. Specifically, letting $\alpha = k_f p_{tot}$ and $K = \sqrt{K_m K_d}$, we have that

$$F(X) = \alpha\frac{(X/K)^2}{1 + (X/K)^2}$$

is the standard Hill function expression. $\triangledown$

Example 3.12 (Enzymatic reaction). Recall the enzymatic reaction

$$E + S \underset{d}{\overset{a}{\rightleftharpoons}} C \overset{k}{\to} E + P,$$

in which E is an enzyme, S is the substrate to which the enzyme binds to form the complex C, and P is the product resulting from the modification of the substrate S due to the binding with the enzyme E. The corresponding system of differential equations is given by

$$\frac{dE}{dt} = -aES + dC + kC, \qquad \frac{dC}{dt} = aES - (d + k)C,$$

$$\frac{dS}{dt} = -aES + dC, \qquad\qquad \frac{dP}{dt} = kC. \tag{3.25}$$

By considering that binding and unbinding reactions are much faster than the catalytic reactions, mathematically expressed by $d \gg k$, we showed before that by approximating $dC/dt = 0$, we obtain $C = E_{tot}S/(S + K_m)$, with $K_m = (d+k)/a$ and $dP/dt = V_{max}S/(S + K_m)$ with $V_{max} = kE_{tot}$. From this, it also follows that

$$\frac{dE}{dt} \approx 0 \qquad \text{and} \qquad \frac{dS}{dt} \approx -\frac{dP}{dt}. \qquad (3.26)$$

How good is this approximation? By applying the singular perturbation method, we will obtain a clear answer to this question. Specifically, define $K_d := d/a$ and convert the system to standard singular perturbation form by defining the small parameter $\epsilon := k/d$, so that $d = k/\epsilon$, $a = k/(K_d\epsilon)$, and the system becomes

$$\epsilon\frac{dE}{dt} = -\frac{k}{K_d}ES + kC + \epsilon kC, \qquad \epsilon\frac{dC}{dt} = \frac{k}{K_d}ES - kC - \epsilon kC,$$
$$\epsilon\frac{dS}{dt} = -\frac{k}{K_d}ES + kC, \qquad \frac{dP}{dt} = kC.$$

We cannot directly apply singular perturbation theory on this system because from the linearization of the first three equations, we see that the boundary layer dynamics are not locally asymptotically stable since there are two zero eigenvalues. This is because the three variables E, S, C are not independent. Specifically, $E = E_{tot} - C$ and $S + C + P = S(0) = S_{tot}$, assuming that initially we have S in amount $S(0)$ and no P and C in the system. Given these conservation laws, the system can be rewritten as

$$\epsilon\frac{dC}{dt} = \frac{k}{K_d}(E_{tot} - C)(S_{tot} - C - P) - kC - \epsilon kC, \qquad \frac{dP}{dt} = kC.$$

Under the assumption made in the analysis of the enzymatic reaction that $S_{tot} \gg E_{tot}$, we have that $C \ll S_{tot}$ so that the equations finally become

$$\epsilon\frac{dC}{dt} = \frac{k}{K_d}(E_{tot} - C)(S_{tot} - P) - kC - \epsilon kC, \qquad \frac{dP}{dt} = kC.$$

We can verify (see Exercise 3.11) that in this system the boundary layer dynamics are locally asymptotically stable, so that setting $\epsilon = 0$ one obtains

$$\bar{C} = \frac{E_{tot}(S_{tot} - \bar{P})}{(S_{tot} - \bar{P}) + K_m} =: h(\bar{P}),$$

and thus that the reduced system is given by

$$\frac{d\bar{P}}{dt} = V_{max}\frac{(S_{tot} - \bar{P})}{(S_{tot} - \bar{P}) + K_m}.$$

This system is the same as that obtained in Chapter 2. However, $dC(t)/dt$ and $dE(t)/dt$ are not close to zero as obtained earlier. In fact, from the conservation

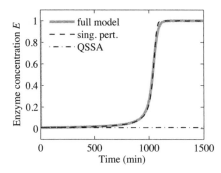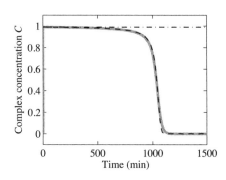

Figure 3.22: Simulation results for the enzymatic reaction comparing the approximations from singular perturbation and from the quasi-steady state approximation (QSSA). Here, we have $S_{\text{tot}} = 100$ nM, $E_{\text{tot}} = 1$ nM, $a = 10$ nM^{-1} min^{-1}, $d = 10$ min^{-1}, and $k = 0.1$ min^{-1}. The full model is the one in equations (3.25).

law $\bar{S} + \bar{C} + \bar{P} = S(0) = S_{\text{tot}}$, we obtain that $d\bar{S}/dt = -d\bar{P}/dt - d\bar{C}/dt$, in which now $d\bar{C}/dt = \partial h/\partial P(\bar{P}) \cdot d\bar{P}/dt$. Therefore, we have that

$$\frac{d\bar{E}}{dt} = -\frac{d\bar{C}}{dt} = -\frac{\partial h}{\partial P}(\bar{P})\frac{d\bar{P}}{dt}, \qquad E(0) = E_{\text{tot}} - h(\bar{P}(0)), \qquad (3.27)$$

and

$$\frac{d\bar{S}}{dt} = -\frac{d\bar{P}}{dt}\left(1 + \frac{\partial h}{\partial P}(\bar{P})\right), \qquad \bar{S}(0) = S_{\text{tot}} - h(\bar{P}(0)) - \bar{P}(0), \qquad (3.28)$$

which are different from expressions (3.26).

These expressions are close to those in equation (3.26) only when $\partial h/\partial P$ is small enough. In the plots of Figure 3.22, we show the time trajectories of the original system, of the Michaelis-Menten quasi-steady state approximation (QSSA), and of the singular perturbation approximation. In the original model (solid line in Figure 3.22), $E(t)$ starts from a unit concentration and immediately collapses to zero as the enzyme is all consumed to form the complex C by the substrate, which is in excess. Similarly, $C(t)$ starts from zero and immediately reaches the maximum possible value of one.

In the quasi-steady state approximation, both $E(t)$ and $C(t)$ are assumed to stabilize immediately to their (quasi) steady state and then stay constant. This is depicted by the dash-dotted plots in Figure 3.22, in which $E(t)$ stays at zero for the whole time and $C(t)$ stays at one for the whole time. This approximation is fairly good as long as there is an excess of substrate. When the substrate concentration goes to zero as it is all converted to product, the complex concentration C goes back to zero (see solid line of Figure 3.22). At this time, the concentrations of complex and enzyme substantially change with time and the quasi-steady state approximation is unsatisfactory. By contrast, the reduced dynamics obtained from the singular perturbation approach well represent the dynamics of the full system even during

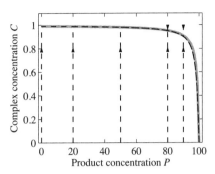

Figure 3.23: The slow manifold of the system $C = h(P)$ is shown by the solid line. The dashed lines show the trajectories of the full system (3.25). These trajectories collapse into an ϵ-neighbor of the slow manifold.

this transient time. Hence, while the quasi-steady state approximation is good only as long as there is an excess of substrate in the system, the reduced dynamics obtained by the singular perturbation approach are a good approximation even when the substrate concentration goes to zero.

In Figure 3.23, we show the curve $C = h(P)$ and the trajectories of the full system. All of the trajectories of the system immediately collapse into an ϵ-neighbor of the curve $C = h(P)$. From this plot, it is clear that $\partial h / \partial P$ is small as long as the product concentration P is small enough, which corresponds to a substrate concentration S large enough. This confirms that the quasi-steady state approximation is good only as long as there is excess of substrate S. $\qquad \nabla$

Exercises

3.1 (Frequency response of a phosphorylation cycle) Consider the model of a covalent modification cycle as illustrated in Chapter 2 in which the kinase Z is not conserved, but it is produced and decays according to the reaction $Z \underset{u(t)}{\overset{\gamma}{\rightleftharpoons}} \emptyset$. Let $u(t)$ be the input stimulus of the cycle and let X^* be the output. Determine the frequency response of X^* to u, determine its bandwidth, and make plots of it. What parameters can be used to tune the bandwidth?

3.2 (Design for robustness) Consider a one-step reaction model for a phosphorylation cycle as seen in Section 2.4, in which the input stimulus is the time-varying concentration of kinase $Z(t)$. When found in the cellular environment, this cycle is subject to possible interactions with other cellular components, such as the non-specific or specific binding of X^* to target sites, to noise due to stochasticity of the cellular environment, and to other crosstalk phenomena. For now, we can think of these disturbances as acting like an aggregate rate of change on the output protein X^*, which we call $d(t)$. Hence, we can model the "perturbed" cycle by

$$\frac{X^*}{dt} = Z(t)k_1 X_{\text{tot}}\left(1 - \frac{X^*}{X_{\text{tot}}}\right) - k_2 Y_{\text{tot}} X^* + d(t).$$

Assume that you can tune all the parameters in this system. Can you tune these parameters so that the response of $X^*(t)$ to $d(t)$ is arbitrarily attenuated while the response of $X^*(t)$ to $Z(t)$ remains arbitrarily large? If yes, explain how these parameters should be tuned to reach this design objective.

3.3 (Design limitations) This problem illustrates possible limitations that are involved in any realistic design question. Here, we examine this through the open loop and negative feedback transcriptional component. Specifically, we want to compare the robustness of these two topologies to perturbations. We model these perturbations as a time-varying disturbance affecting the production rate of mRNA m and protein P. To slightly simplify the problem, we focus only on disturbances affecting the production of protein. The open loop model becomes

$$\frac{dm_{\text{P}}}{dt} = \alpha_0 - \delta m_{\text{P}}, \qquad \frac{dP}{dt} = \kappa m_{\text{P}} - \gamma P + d(t),$$

and the negative feedback system becomes

$$\frac{dm_{\text{P}}}{dt} = \alpha_0 + \frac{\alpha}{1 + (P/K)^n} - \delta m_{\text{P}}, \qquad \frac{dP}{dt} = \kappa m_{\text{P}} - \gamma P + d(t).$$

Answer the following questions:

(i) After performing linearization about the equilibrium point, determine analytically the frequency response of P to d for both systems.

(ii) Sketch the magnitude plot of this response for both systems, compare them, and determine what happens as κ and α increase (note: if your calculations are correct, you should find that what really matters for the negative feedback system is the product $\alpha\kappa$, which we can view as the *feedback gain*). Is increasing the feedback gain the best strategy to decrease the sensitivity of the system to the disturbance?

(iii) Pick parameter values and use MATLAB to draw plots of the frequency response magnitude and phase as the feedback gain increases and validate your predictions in (b). (Suggested parameters: $\delta = 1$ hrs^{-1}, $\gamma = 1$ hrs^{-1}, $K = 1$ nM, $n = 1$, $\alpha\kappa = \{1, 10, 100, 1000, \ldots\}$.)

(iv) Investigate the answer to (c) when you have $\delta = 20$ hrs^{-1}, that is, the timescale of the mRNA dynamics becomes faster than that of the protein dynamics. What changes with respect to what you found in (c)?

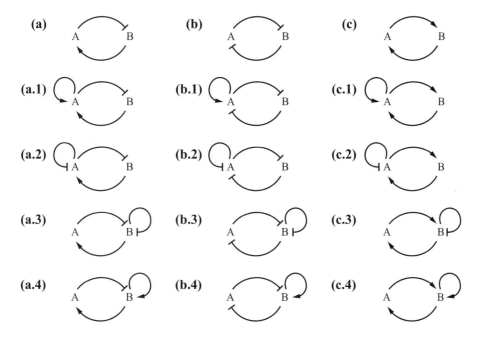

Figure 3.24: Circuit topologies with two proteins: A and B.

(v) When δ is at least 10 times larger than γ, you can approximate the m dynamics to the quasi-steady state. So, the two above systems can be reduced to one differential equation. For these two reduced systems, determine analytically the frequency response to d and use it to determine whether arbitrarily increasing the feedback gain is a good strategy to decrease the sensitivity of response to the disturbance.

3.4 (Adaptation) Show that the dynamics of the "sniffer" in equation (3.16) can be taken into the standard integral feedback form through a suitable change of coordinates.

3.5 (Bendixson criterion) Consider the system

$$\frac{dx_1}{dt} = -x_2^3 + \delta x_1^3, \qquad \frac{dx_2}{dt} = x_1^3.$$

When $\delta > 0$, Bendixson's criterion rules out the existence of a periodic solution in $\mathbb{R}^2$. Assume now $\delta = 0$, and determine whether the system admits a limit cycle in $\mathbb{R}^2$. (Hint: consider the function $V = x_1^2 + x_2^2$ and determine the behavior of $V(t)$ when $x_1(t)$ and $x_2(t)$ are solutions to the above system.)

3.6 (Bendixson criterion) Consider the possible circuit topologies of Figure 3.24, in which A and B are proteins and activation ($\rightarrow$) and repression ($\dashv$) interactions represent transcriptional activation or repression. Approximate the mRNA dynamics at the quasi-steady state. Use Bendixson's criterion to rule out topologies that cannot give rise to closed orbits.

3.7 (Two gene oscillator) Consider the feedback system composed of two genes expressing proteins A (activator) and R (repressor), in which we denote by A, R, m_A, and m_R, the concentrations of the activator protein, the repressor protein, the mRNA for the activator protein, and the mRNA for the repressor protein, respectively. The differential equation model corresponding to this system is given by

$$\frac{dm_A}{dt} = \frac{\alpha}{1 + (R/K_1)^n} - \delta m_A, \qquad \frac{dm_R}{dt} = \frac{\alpha(A/K_2)^m}{1 + (A/K_2)^m} - \delta m_R,$$

$$\frac{dA}{dt} = \kappa m_A - \gamma A, \qquad\qquad \frac{dR}{dt} = \kappa m_R - \gamma R.$$

Determine parameter conditions under which this system admits a stable limit cycle. Validate your findings through simulation.

3.8 (Goodwin oscillator) Consider the simple set of reactions

$$X_1 \xrightarrow{k} X_2 \xrightarrow{k} X_3 ... \xrightarrow{k} X_n.$$

Assume further that X_n is a transcription factor that represses the production of protein X_1 through transcriptional regulation (assume simple binding of X_n to DNA). Neglecting the mRNA dynamics of X_1, write the differential equation model of this system and determine conditions on the length n of the cascade for which the system admits a stable limit cycle. Validate your findings through simulation.

3.9 (Phosphorylation via singular perturbation) Consider again the model of a covalent modification cycle as illustrated in Section 2.4 in which the kinase Z is not constant, but it is produced and decays according to the reaction

$$Z \underset{u(t)}{\overset{\gamma}{\rightleftharpoons}} \emptyset.$$

(i) Consider that $d \gg k, \gamma, u(t)$ and employ singular perturbation with small parameter $\epsilon = \gamma/d$ to obtain the approximated dynamics of $Z(t)$ and $X^*(t)$. How is this different from the result obtained in Exercise 2.12?

(ii) Simulate these approximated dynamics when $u(t)$ is a periodic signal with frequency ω and compare the responses of Z of these approximated dynamics to those obtained in Exercise 2.12 as you change ω. What do you observe? Explain.

3.10 (Hill function via singular perturbation) Show that the slow manifold of the following system is asymptotically stable:

$$\epsilon\frac{dX_2}{dt} = c\frac{\gamma}{K_m}X^2 - c\gamma X_2 - \frac{\gamma}{K_d}X_2(p_{tot} - C) + \gamma C, \qquad \frac{dm_Y}{dt} = \alpha C - \delta m_Y,$$

$$\epsilon\frac{dC}{dt} = \frac{\gamma}{K_d}X_2(p_{tot} - C) - \gamma C, \qquad \frac{dY}{dt} = \beta m_Y - \gamma Y.$$

3.11 (Enzyme dynamics via singular perturbation) Show that the slow manifold of the following system is asymptotically stable:

$$\epsilon\frac{dC}{dt} = \frac{k}{K_d}(E_{tot} - C)\cdot(S_{tot} - P) - kC - \epsilon kC, \qquad \frac{dP}{dt} = kC.$$

Chapter 4
Stochastic Modeling and Analysis

In this chapter we explore stochastic behavior in biomolecular systems, building on our preliminary discussion of stochastic modeling in Section 2.1. We begin by reviewing methods for modeling stochastic processes, including the chemical master equation (CME), the chemical Langevin equation (CLE) and the Fokker-Planck equation (FPE). Given a stochastic description, we can then analyze the behavior of the system using a collection of stochastic simulation and analysis tools. This chapter makes use of a variety of topics in stochastic processes; readers should have a good working knowledge of basic probability and some exposure to simple stochastic processes.

4.1   Stochastic modeling of biochemical systems

Biomolecular systems are inherently noisy due to the random nature of molecular reactions. When the concentrations of molecules are high, the deterministic models we have used in the previous chapters provide a good description of the dynamics of the system. However, if the molecular counts are low then it is often necessary to explicitly account for the random nature of events. In this case, the chemical reactions in the cell can be modeled as a collection of stochastic events corresponding to chemical reactions between species. These include binding and unbinding of molecules (such as RNA polymerase and DNA), conversion of one set of species into another, and enzymatically controlled covalent modifications such as phosphorylation. In this section we will briefly survey some of the different representations that can be used for stochastic models of biochemical systems, following the material in the textbooks by Phillips et al. [78], Gillespie [32] and Van Kampen [53].

Statistical mechanics

At the core of many of the reactions and multi-molecular interactions that take place inside of cells is the chemical physics associated with binding between two molecules. One way to capture some of the properties of these interactions is through the use of statistical mechanics and thermodynamics.

As described briefly already in Chapter 2, the underlying representation for both statistical mechanics and chemical kinetics is to identify the appropriate mi-

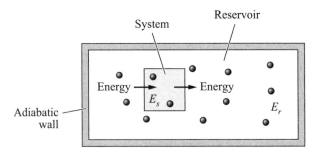

Figure 4.1: System in contact with a reservoir. While there is exchange of energy between the system and the reservoir, there is no exchange of energy between them and the rest of the world. Figure adapted from [78].

crostates of the system. A microstate corresponds to a given configuration of the components (species) in the system relative to each other and we must enumerate all possible configurations between the molecules that are being modeled.

In statistical mechanics, we model the configuration of the cell by the probability that the system is in a given microstate. This probability can be calculated based on the energy levels of the different microstates. Consider a setting in which our system is in contact with a reservoir (Figure 4.1). Let E_r represent the energy in the reservoir, E_s the energy in the system and $E_{\text{tot}} = E_r + E_s$ the total (conserved) energy. Given two different energy levels E_{q_1} and E_{q_2} for the system of interest, let $W_r(E_{\text{tot}} - E_{q_i})$ be the number of possible microstates of the reservoir with energy $E_r = E_{\text{tot}} - E_{q_i}$, $i = 1, 2$. The laws of statistical mechanics state that the ratio of probabilities of being in microstates q_1 and q_2 is given by the ratio of the number of possible states of the reservoir:

$$\frac{\mathbb{P}(E_{q_1})}{\mathbb{P}(E_{q_2})} = \frac{W_r(E_{\text{tot}} - E_{q_1})}{W_r(E_{\text{tot}} - E_{q_2})}. \tag{4.1}$$

Defining the entropy of the reservoir as $S_r = k_B \ln W_r$, where k_B is Boltzmann's constant, we can rewrite equation (4.1) as

$$\frac{W_r(E_{\text{tot}} - E_{q_1})}{W_r(E_{\text{tot}} - E_{q_2})} = \frac{e^{S_r(E_{\text{tot}} - E_{q_1})/k_B}}{e^{S_r(E_{\text{tot}} - E_{q_2})/k_B}}.$$

We now approximate $S_r(E_{\text{tot}} - E_s)$ in a Taylor series expansion around E_{tot}, under the assumption that $E_r \gg E_{q_i}$:

$$S_r(E_{\text{tot}} - E_s) \approx S_r(E_{\text{tot}}) - \frac{\partial S_r}{\partial E} E_s.$$

From the properties of thermodynamics, if we hold the volume and number of molecules constant, then we can define the temperature as

$$\left. \frac{\partial S}{\partial E} \right|_{V,N} = \frac{1}{T}$$

and we obtain

$$\frac{\mathbb{P}(E_{q_1})}{\mathbb{P}(E_{q_2})} = \frac{e^{-E_{q_1}/k_B T}}{e^{-E_{q_2}/k_B T}}.$$

This implies that

$$\mathbb{P}E_q \propto e^{-E_q/(k_B T)}$$

and hence the probability of being in a microstate q is given by

$$\mathbb{P}(q) = \frac{1}{Z} e^{-E_q/(k_B T)}, \tag{4.2}$$

where we have written E_q for the energy of the microstate and Z is a normalizing factor, known as the *partition function*, defined by

$$Z = \sum_{q \in Q} e^{-E_q/(k_B T)}.$$

In many situations we do not care about the specific microstate that a system is in, but rather whether the system is in any one of a number of microstates that all correspond to the same overall behavior of the system. For example, we will often not care whether a specific RNA polymerase is bound to a promoter, but rather whether *any* RNA polymerase is bound to that promoter. We call the collection of microstates that is of interest a *macrostate* (or sometimes *system state*). A macrostate is defined as a set of states $S \subset Q$ that correspond to a given condition that we wish to monitor. Given a macrostate S, the probability of being in that macrostate is

$$\mathbb{P}(S) = \frac{1}{Z} \sum_{q \in S} e^{-E_q/(k_B T)} = \frac{\sum_{q \in S} e^{-E_q/(k_B T)}}{\sum_{q \in Q} e^{-E_q/(k_B T)}}. \tag{4.3}$$

It is this probability that allows us, for example, to determine whether any RNA polymerase molecule is bound to a given promoter, averaged over many independent samples. We can then use this probability to determine the rate of expression of the corresponding gene.

Example 4.1 (Transcription factor binding). Suppose that we have a transcription factor R that binds to a specific target region on a DNA strand (such as the promoter region upstream of a gene). We wish to find the probability P_{bound} that the transcription factor will be bound to this location as a function of the number of transcription factor molecules n_R in the system. If the transcription factor is a repressor, for example, knowing $P_{bound}(n_R)$ will allow us to calculate the likelihood of transcription occurring.

To compute the probability of binding, we assume that the transcription factor can bind non-specifically to other sections of the DNA (or other locations in the cell) and we let N_{ns} represent the number of such sites. We let E_{bound} represent the free energy associated with R bound to its specified target region and E_{ns} represent the free energy for R in any other non-specific location, where we assume that $E_{\text{bound}} < E_{\text{ns}}$. The microstates of the system consist of all possible assignments of the n_R transcription factors to either a non-specific location or the target region of the DNA. Since there is only one target site, there can be at most one transcription factor attached there and hence we must count all of the ways in which either zero or one molecule of R are attached to the target site.

If none of the n_R copies of R are bound to the target region then these must be distributed between the N_{ns} non-specific locations. Each bound protein has energy E_{ns}, so the total energy for any such configuration is $n_R E_{\text{ns}}$. The number of such combinations is $\binom{N_{\text{ns}}}{n_R}$, assuming the R's are indistinguishable, and so the contribution to the partition function from these microstates is

$$Z_{\text{ns}} = \binom{N_{\text{ns}}}{n_R} e^{-n_R E_{\text{ns}}/(k_B T)} = \frac{N_{\text{ns}}!}{n_R!(N_{\text{ns}} - n_R)!} e^{-n_R E_{\text{ns}}/(k_B T)}.$$

For the microstates in which one molecule of R is bound at a target site and the other $n_R - 1$ molecules are at the non-specific locations, we have a total energy of $E_{\text{bound}} + (n_R - 1)E_{\text{ns}}$ and $\binom{N_{\text{ns}}}{(n_R-1)}$ possible such states. The resulting contribution to the partition function is

$$Z_{\text{bound}} = \frac{N_{\text{ns}}!}{(n_R - 1)!(N_{\text{ns}} - n_R + 1)!} e^{-(E_{\text{bound}} - (n_R-1)E_{\text{ns}})/(k_B T)}.$$

The probability that the target site is occupied is now computed by looking at the ratio of the Z_{bound} to $Z = Z_{\text{ns}} + Z_{\text{bound}}$. After some basic algebraic manipulations, it can be shown that

$$P_{\text{bound}}(n_R) = \frac{\left(\frac{n_R}{N_{\text{ns}}-n_R+1}\right)\exp[-(E_{\text{bound}} + E_{\text{ns}})/(k_B T)]}{1 + \left(\frac{n_R}{N_{\text{ns}}-n_R+1}\right)\exp[-(E_{\text{bound}} + E_{\text{ns}})/(k_B T)]}.$$

If we assume that $N_{\text{ns}} \gg n_R$ then $N_{\text{ns}} - n_R + 1 \approx N_{\text{ns}}$, and we can write

$$P_{\text{bound}}(n_R) \approx \frac{kn_R}{1 + kn_R}, \quad \text{where} \quad k = \frac{1}{N_{\text{ns}}}\exp[-(E_{\text{bound}} - E_{\text{ns}})/(k_B T)]. \qquad (4.4)$$

As we would expect, this says that for very small numbers of repressors, P_{bound} is close to zero, while for large numbers of repressors, $P_{\text{bound}} \to 1$. The point at which we get a binding probability of 0.5 is when $n_R = 1/k$, which depends on the relative binding energies and the number of non-specific binding sites. $\qquad \nabla$

Example 4.2 (Combinatorial promoter). As mentioned in Section 2.3, a combinatorial promoter is a region of DNA in which multiple transcription factors can bind and influence the subsequent binding of RNA polymerase (RNAP). Combinatorial promoters appear in a number of natural and engineered circuits and represent a mechanism for creating switch-like behavior.

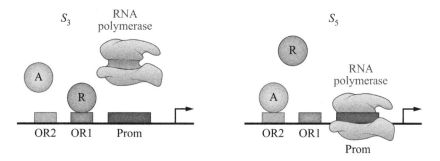

Figure 4.2: Two possible configurations of a combinatorial promoter where both an activator and a repressor can bind to specific operator sites. We show configurations S_3 and S_5 referring to Table 4.1.

One method to model a combinatorial promoter is to use the binding energies of the different combinations of proteins to the operator region, and then compute the probability of being in a given promoter state given the concentration of each of the transcription factors. Table 4.1 shows the possible states of a notional promoter that has two operator regions—one that binds a repressor protein R and another that binds an activator protein A.

As indicated in the table, the promoter has three (possibly overlapping) regions of DNA: OR1 and OR2 are binding sites for the repressor and activator proteins, respectively, and Prom is the location where RNA polymerase binds. (The individual labels are primarily for bookkeeping purposes and may not correspond to physically separate regions of DNA.)

To determine the probabilities of being in a given macrostate, we must compute the individual microstates that occur at given concentrations of repressor, activator and RNA polymerase. Each microstate corresponds to an individual set of molecules binding in a specific configuration. So if we have n_R repressor molecules,

Table 4.1: Configurations for a combinatorial promoter with an activator and a repressor. Each row corresponds to a specific macrostate of the promoter in which the listed molecules are bound to the target region. The relative energy of a state compared with the ground state provides a measure of the likelihood of that state occurring, with more negative numbers corresponding to more energetically favorable configurations.

State S_q	OR1	OR2	Prom	E_q (ΔG)	Comment
S_1	–	–	–	0	No binding (ground state)
S_2	–	–	RNAP	−5	RNA polymerase bound
S_3	R	–	–	−10	Repressor bound
S_4	–	A	–	−12	Activator bound
S_5	–	A	RNAP	−15	Activator and RNA polymerase

then there is one microstate corresponding to *each* different repressor molecule that is bound, resulting in n_R individual microstates. In the case of configuration S_5, where two different molecules are bound, the number of combinations is given by the product of the numbers of individual molecules, $n_A \cdot n_{RNAP}$, reflecting the possible combinations of molecules that can occupy the promoter sites. The overall partition function is given by summing up the contributions from each microstate:

$$Z = e^{-E_1/(k_B T)} + n_{RNAP} e^{-E_2/(k_B T)} + n_R e^{-E_3/(k_B T)}$$
$$+ n_A e^{-E_4/(k_B T)} + n_A n_{RNAP} e^{-E_5/(k_B T)}. \quad (4.5)$$

The probability of a given macrostate is determined using equation (4.3). For example, if we define the promoter to be "active" if RNA polymerase is bound to the DNA, then the probability of being in this macrostate as a function of the various molecular counts is given by

$$P_{active}(n_R, n_A, n_{RNAP}) = \frac{1}{Z}\left(n_{RNAP} e^{-E_2/(k_B T)} + n_A n_{RNAP} e^{-E_5/(k_B T)}\right)$$
$$= \frac{k_5 n_A + k_2}{1 + k_2 + k_3 n_R + (k_4 + k_5)n_A},$$

where

$$k_q = e^{-(E_q - E_1)/(k_B T)}.$$

From this expression we see that if $n_R \gg n_A$ then P_{active} tends to 0 while if $n_A \gg n_R$ then P_{active} tends to 1, as expected.

$\triangledown$

Chemical master equation (CME)

The statistical physics model we have just considered gives a description of the *steady state* properties of the system. In many cases, it is clear that the system reaches this steady state quickly and hence we can reason about the behavior of the system just by modeling the energy of the system. In other situations, however, we care about the transient behavior of a system or the dynamics of a system that does not have an equilibrium configuration. In these instances, we must extend our formulation to keep track of how quickly the system transitions from one microstate to another, known as the *chemical kinetics* of the system.

To model these dynamics, we return to our enumeration of all possible microstates of the system. Let $P(q, t)$ represent the probability that the system is in microstate q at a given time t. Here q can be any of the very large number of possible microstates for the system, which for chemical reaction systems we can represent in terms of a vector consisting of the number of molecules of each species that is present. We wish to write an explicit expression for how $P(q, t)$ varies as a function of time, from which we can study the stochastic dynamics of the system.

We begin by assuming we have a set of M reactions Rj, $j = 1, \ldots, M$, with ξ_j representing the change in state associated with reaction Rj. Specifically, ξ_j is given by the jth column of the stoichiometry matrix N (Section 2.1). The *propensity function* defines the probability that a given reaction occurs in a sufficiently small time step dt:

$$a_j(q,t)dt \quad = \quad \text{Probability that reaction Rj will occur between time } t$$
$$\text{and time } t + dt \text{ given that the microstate is } q.$$

The linear dependence on dt relies on the fact that dt is chosen sufficiently small. We will typically assume that a_j does not depend on the time t and write $a_j(q)dt$ for the probability that reaction j occurs in state q.

Using the propensity function, we can compute the distribution of states at time $t + dt$ given the distribution at time t:

$$P(q, t+dt) = P(q,t) \prod_{j=1}^{M} \left(1 - a_j(q)dt \right) + \sum_{j=1}^{M} P(q - \xi_j) a_j(q - \xi_j) dt$$

$$= P(q,t) + \sum_{j=1}^{M} \left(a_j(q - \xi_j) P(q - \xi_j, t) - a_j(q) P(q,t) \right) dt + O(dt^2), \tag{4.6}$$

where $O(dt^2)$ represents higher order terms in dt. Since dt is small, we can take the limit as $dt \to 0$ and we obtain the *chemical master equation* (CME):

$$\frac{\partial P}{\partial t}(q,t) = \sum_{j=1}^{M} \left(a_j(q - \xi_j) P(q - \xi_j, t) - a_j(q) P(q,t) \right). \tag{4.7}$$

This equation is also referred to as the *forward Kolmogorov equation* for a discrete state, continuous time random process.

Despite its complexity, the master equation does capture many of the important details of the chemical physics of the system and we shall use it as our basic representation of the underlying dynamics. As we shall see, starting from this equation we can then derive a variety of alternative approximations that allow us to answer specific questions of interest.

The key element of the master equation is the propensity function $a_j(q)$, which governs the rate of transition between microstates. Although the detailed value of the propensity function can be quite complex, its functional form is often relatively simple. In particular, for a unimolecular reaction of the form A $\to$ B, the propensity function is proportional to the number of molecules of A that are present:

$$a_j(q) = k_j n_A. \tag{4.8}$$

This follows from the fact that the reaction associated with each molecule is independent and hence the likelihood of a reaction happening depends directly on the number of copies of A that are present.

Similarly, for a bimolecular reaction, we have that the likelihood of a reaction occurring is proportional to the product of the number of molecules of each type that are present (since this is the number of independent reactions that can occur) and inversely proportional to the volume Ω. Hence, for a reaction of the form A + B $\longrightarrow$ C we have

$$a_j(q) = \frac{k_j}{\Omega} n_A n_B. \tag{4.9}$$

The rigorous verification of this functional form is beyond the scope of this text, but roughly we keep track of the likelihood of a single reaction occurring between A and B and then multiply by the total number of combinations of the two molecules that can react ($n_A \cdot n_B$).

A special case of a bimolecular reaction occurs when A = B, so that our reaction is given by A + A $\rightarrow$ B. In this case we must take into account that a molecule cannot react with itself and that the molecules are indistinguishable, and so the propensity function is of the form

$$a_j(q) = \frac{k_j}{\Omega} \cdot \frac{n_A(n_A - 1)}{2}. \tag{4.10}$$

Here, $n_A(n_A - 1)/2$ represents the number of ways that two molecules can be chosen from a collection of n_A identical molecules.

Note that the use of the parameter k_j in the propensity functions above is intentional since it corresponds to the reaction rate parameter that is present in the reaction rate equation models we used in Chapter 2. The factor of Ω for bimolecular reactions models the fact that the propensity of a bimolecular reaction occurring depends explicitly on the volume in which the reaction takes place.

Although it is tempting to extend the formula for a bimolecular reaction to the case of more than two species being involved in a reaction, usually such reactions actually involve combinations of bimolecular reactions, e.g.:

$$A + B + C \longrightarrow D \qquad \Longrightarrow \qquad A + B \longrightarrow AB, \quad AB + C \longrightarrow D.$$

This more detailed description reflects the fact that it is extremely unlikely that three molecules will all come together at precisely the same instant. The much more likely possibility is that two molecules will initially react, followed by a second reaction involving the third molecule.

Example 4.3 (Repression of gene expression). We consider a simple model of repression in which we have a promoter that contains binding sites for RNA polymerase and a repressor protein R. RNA polymerase only binds when the repressor is absent, after which it can undergo an isomerization reaction to form an open complex and initiate transcription (see Section 2.2). Once the RNA polymerase begins to create mRNA, we assume the promoter region is uncovered, allowing another repressor or RNA polymerase to bind.

The following reactions describe this process:

R1: $R + DNA \longrightarrow R{:}DNA,$

R2: $R{:}DNA \longrightarrow R + DNA,$

R3: $RNAP + DNA \longrightarrow RNAP{:}DNA^c,$

R4: $RNAP{:}DNA^c \longrightarrow RNAP + DNA,$

R5: $RNAP{:}DNA^c \longrightarrow RNAP{:}DNA^o,$

R6: $RNAP{:}DNA^o \longrightarrow RNAP + DNA + mRNA,$

where $RNAP : DNA^c$ represents the closed complex and $RNAP : DNA^o$ represents the open complex, and reaction R6 lumps together start of transcription, elongation, mRNA creation, and termination. The states for the system depend on the number of molecules of each species that are present. If we assume that we start with n_R repressors and n_{RNAP} RNA polymerases, then the possible states (S) for our system are outlined below.

S	DNA	R	RNAP	R : DNA	RNAP : DNA^c	RNAP : DNA^o
q_1	1	n_R	n_{RNAP}	0	0	0
q_2	0	$n_R - 1$	n_{RNAP}	1	0	0
q_3	0	n_R	$n_{RNAP} - 1$	0	1	0
q_4	0	n_R	$n_{RNAP} - 1$	0	0	1

Note that we do not keep track of each individual repressor or RNA polymerase molecule that binds to the DNA, but simply keep track of whether they are bound or not.

We can now rewrite the chemical reactions as a set of transitions between the possible microstates of the system. Assuming that all reactions take place in a volume Ω, we use the propensity functions for unimolecular and bimolecular reactions to obtain:

R1: $q_1 \longrightarrow q_2;$ $a_1(q_1) = (k_1/\Omega)n_R,$ R4: $q_3 \longrightarrow q_1;$ $a_4(q_3) = k_4,$

R2: $q_2 \longrightarrow q_1;$ $a_2(q_2) = k_2,$ R5: $q_3 \longrightarrow q_4;$ $a_5(q_3) = k_5,$

R3: $q_1 \longrightarrow q_3;$ $a_3(q_1) = (k_3/\Omega)n_{RNAP},$ R6: $q_4 \longrightarrow q_1;$ $a_6(q_4) = k_6.$

The chemical master equation can now be written down using the propensity functions for each reaction:

$$\frac{d}{dt}\begin{pmatrix} P(q_1,t) \\ P(q_2,t) \\ P(q_3,t) \\ P(q_4,t) \end{pmatrix} = \begin{pmatrix} -(k_1/\Omega)n_R - (k_3/\Omega)n_{RNAP} & k_2 & k_4 & k_6 \\ (k_1/\Omega)n_R & -k_2 & 0 & 0 \\ (k_3/\Omega)n_{RNAP} & 0 & -k_4 - k_5 & 0 \\ 0 & 0 & k_5 & -k_6 \end{pmatrix} \begin{pmatrix} P(q_1,t) \\ P(q_2,t) \\ P(q_3,t) \\ P(q_4,t) \end{pmatrix}.$$

The initial condition for the system can be taken as $P(q,0) = (1,0,0,0)$, corresponding to the state q_1. A simulation showing the evolution of the probabilities is shown in Figure 4.3.

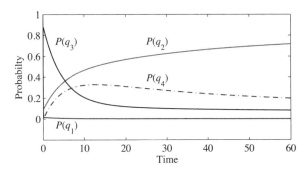

Figure 4.3: Numerical solution of chemical master equation for simple repression model.

The equilibrium solution for the probabilities can be solved by setting $\dot{P} = 0$, which yields:

$$P_e(q_1) = \frac{k_2 k_6 \Omega (k_4 + k_5)}{k_1 k_6 n_R (k_4 + k_5) + k_2 k_3 n_{RNAP}(k_5 + k_6) + k_2 k_6 \Omega (k_4 + k_5)},$$

$$P_e(q_2) = \frac{k_1 k_6 n_R (k_4 + k_5)}{k_1 k_6 n_R (k_4 + k_5) + k_2 k_3 n_{RNAP}(k_5 + k_6) + k_2 k_6 \Omega (k_4 + k_5)},$$

$$P_e(q_3) = \frac{k_2 k_3 k_6 n_{RNAP}}{k_1 k_6 n_R (k_4 + k_5) + k_2 k_3 n_{RNAP}(k_5 + k_6) + k_2 k_6 \Omega (k_4 + k_5)},$$

$$P_e(q_4) = \frac{k_2 k_3 k_5 n_{RNAP}}{k_1 k_6 n_R (k_4 + k_5) + k_2 k_3 n_{RNAP}(k_5 + k_6) + k_2 k_6 \Omega (k_4 + k_5)}.$$

We see that the equilibrium distributions depend on the relative strengths of different combinations of the rate constants for the individual reactions in the system. For example, the probability that a repressor molecule is bound to the promoter is given by

$$P_{bound}(n_R) = P_e(q_2) = \frac{k_1 k_6 n_R (k_4 + k_5)}{k_1 k_6 n_R (k_4 + k_5) + k_2 k_3 n_{RNAP}(k_5 + k_6) + k_2 k_6 \Omega (k_4 + k_5)},$$

which has a functional form similar to equation (4.4). Note that here the probability depends on the volume Ω because we used a different model for the diffusion of the repressor R (previously we assumed all repressors were non-specifically bound to DNA).

∇

Example 4.4 (Transcription of mRNA). Consider the production of mRNA from a single copy of DNA. We have two basic reactions that can occur: mRNA can be produced by RNA polymerase transcribing the DNA and producing an mRNA strand, or mRNA can be degraded. We represent the microstate q of the system in terms of the number of mRNA's that are present, which we write as n for ease of notation. The reactions can now be represented as $\xi_1 = +1$, corresponding to transcription, and $\xi_2 = -1$, corresponding to degradation. We choose as our propensity functions

$$a_1(n) = \alpha, \qquad a_2(n) = \delta n,$$

by which we mean that the probability that a gene is transcribed in time dt is αdt and the probability that a transcript is created in time dt is $\delta n dt$ (proportional to the number of mRNA's).

We can now write down the master equation as described above. Equation (4.6) becomes

$$P(n, t+dt) = P(n,t)\left(1 - \sum_{i=1,2} a_i(n)dt\right) + \sum_{i=1,2} P(n - \xi_i, t)a_i(q - \xi_i)dt$$

$$= P(n,t) - a_1(n)P(n,t) - a_2(n)P(n,t)$$

$$+ a_1(n-1)P(n-1,t) + a_2(n+1)P(n+1,t)$$

$$= P(n,t) + \alpha P(n-1,t)dt - (\alpha + \delta n)P(n,t)dt + \delta(n+1)P(n+1,t)dt.$$

This formula holds for $n = 1, 2, \ldots$, with the $n = 0$ case satisfying

$$P(0, t+dt) = P(0,t) - \alpha P(0,t)dt + \delta P(1,t)dt.$$

Notice that we have an infinite number of equations, since n can be any positive integer.

We can write the differential equation version of the master equation by subtracting the first term on the right-hand side and dividing by dt:

$$\frac{d}{dt}P(n,t) = \alpha P(n-1,t) - (\alpha + \delta n)P(n,t) + \delta(n+1)P(n+1,t), \qquad n = 1,2,\ldots$$

$$\frac{d}{dt}P(0,t) = -\alpha P(0,t)dt + \delta P(1,t).$$

Again, this is an infinite number of differential equations, although we could take some limit N and simply declare that $P(N,t) = 0$ to yield a finite number.

One simple type of analysis that can be done on this equation without truncating it to a finite number is to look for a steady state solution to the equation. In this case, we set $\dot{P}(n,t) = 0$ and look for a constant solution $P(n,t) = p_e(n)$. This yields an algebraic set of relations

$$0 = -\alpha p_e(0) + \delta p_e(1) \qquad\qquad \alpha p_e(0) = \delta p_e(1)$$

$$0 = \alpha p_e(0) - (\alpha + \delta)p_e(1) + 2\delta p_e(2) \qquad\qquad \alpha p_e(1) = 2\delta p_e(2)$$

$$0 = \alpha p_e(1) - (\alpha + 2\delta)p_e(2) + 3\delta p_e(3) \qquad\Longrightarrow\qquad \alpha p_e(2) = 3\delta p_e(3)$$

$$\vdots \qquad\qquad\qquad\qquad\qquad\qquad \vdots$$

$$0 = \alpha p_e(n-1) - (\alpha + \delta n)p_e(n) + \delta(n+1)p_e(n+1) \qquad\qquad \alpha p(n-1) = n\delta p(n).$$

Using this recursive expression to obtain $p(n)$ as a function of $p(0)$, we obtain

$$p(n) = \left(\frac{\alpha}{\delta}\right)^n \frac{1}{n!} p(0).$$

Further, using that $\sum_{n=0}^{\infty} p(n) = 1$, we have that

$$\sum_{n=0}^{\infty} \left(\frac{\alpha}{\delta}\right)^n \frac{1}{n!} p(0) = 1,$$

from which, considering that $\sum_{n=0}^{\infty} \left(\frac{\alpha}{\delta}\right)^n \frac{1}{n!} = e^{\alpha/\delta}$, we obtain $p(0) = e^{-\alpha/\delta}$, which finally leads to the Poisson distribution

$$p(n) = e^{\alpha/\delta} \frac{(\alpha/\delta)^n}{n!}.$$

The mean, variance and coefficient of variation (CV), given by the ratio between the standard deviation and the mean, are thus

$$\mu = \frac{\alpha}{\delta}, \qquad \sigma^2 = \frac{\alpha}{\delta}, \qquad CV = \frac{\sigma}{\mu} = \frac{1}{\sqrt{\mu}} = \sqrt{\frac{\delta}{\alpha}}.$$

The coefficient of variation is commonly used to quantify how noisy a process is since it provides a measure of the deviation relative to the mean value. Note that for fixed variance, the coefficient of variation increases if μ decreases. Thus as we have a small number of mRNA molecules present, we see higher variability in the (relative) mRNA concentration. ∇

Chemical Langevin equation (CLE)

The chemical master equation gives a complete description of the evolution of the probability distribution of a system, but it can often be quite cumbersome to work with directly. A number of approximations to the master equation are thus used to provide more tractable formulations of the dynamics. The first of these that we shall consider is known as the *chemical Langevin equation* (CLE).

To derive the chemical Langevin equation, we start by assuming that the number of molecules in the system is large and that we can therefore represent the system using a vector $X \in \mathbb{R}^n$, with X_i representing the (real-valued) number of molecules of species S_i. (Often X_i will be divided by the volume to give a real-valued concentration of species S_i.) In addition, we assume that we are interested in the dynamics on time scales in which individual reactions are not important and so we can look at how the system state changes over time intervals in which many reactions occur and hence the system state evolves in a smooth fashion.

Let $X(t)$ be the state vector for the system, where we assume now that the elements of X are real-valued rather than integer valued. We make the further approximation that we can lump together multiple reactions so that instead of keeping track of the individual reactions, we can average across a number of reactions over a time τ to allow the continuous state to evolve in continuous time. The resulting dynamics can be described by a stochastic process of the form

$$X_i(t + \tau) = X_i(t) + \sum_{j=1}^{M} \xi_{ij} a_j(X(t))\tau + \sum_{j=1}^{M} \xi_{ij} a_j^{1/2}(X(t))N_j(0, \tau),$$

where a_j are the propensity functions for the individual reactions, ξ_{ij} are the corresponding changes in the system states X_i and N_j are a set of independent Gaussian random variables with zero mean and variance τ.

If we assume that τ is small enough that we can use the derivative to approximate the previous equation (but still large enough that we can average over multiple reactions), then we can write

$$\frac{dX_i(t)}{dt} = \sum_{j=1}^{M} \xi_{ji} a_j(X(t)) + \sum_{j=1}^{M} \xi_{ji} a_j^{1/2}(X(t))\Gamma_j(t) =: A_i(X(t)) + \sum_{j=1}^{M} B_{ij}(X(t))\Gamma_j(t),$$

$$(4.11)$$

where Γ_j are white noise processes (see Section 4.3). This equation is called the *chemical Langevin equation* (CLE).

Example 4.5 (Protein production). Consider a simplified two-step model of protein production in which mRNA is produced by DNA and protein by mRNA. We do not model the detailed processes of isomerization and elongation of the mRNA and polypeptide chains. We can capture the state of the system by keeping track of the number of copies of DNA, mRNA, and protein, which we denote by X_D, X_m and X_P, respectively, so that $X = (X_D, X_m, X_P)$.

The simplified reactions with the corresponding propensity functions are given by

$$R1: \quad DNA \xrightarrow{\alpha} mRNA + DNA, \qquad \xi_1 = (0, 1, 0), \qquad a_1(X) = \alpha X_D,$$

$$R2: \quad mRNA \xrightarrow{\delta} \phi, \qquad \xi_2 = (0, -1, 0), \qquad a_2(X) = \delta X_m,$$

$$R3: \quad mRNA \xrightarrow{\kappa} mRNA + protein, \qquad \xi_3 = (0, 0, 1), \qquad a_3(X) = \kappa X_m,$$

$$R4: \quad protein \xrightarrow{\gamma} \phi, \qquad \xi_4 = (0, 0, -1), \qquad a_4(X) = \gamma X_P.$$

Using these, we can write the Langevin equation as

$$\frac{dX_m}{dt} = \alpha X_D - \delta X_m + \sqrt{\alpha X_D}\Gamma_1(t) - \sqrt{\delta X_m}\Gamma_2(t),$$

$$\frac{dX_P}{dt} = \kappa X_m - \gamma X_P + \sqrt{\kappa X_m}\Gamma_3(t) - \sqrt{\gamma X_p}\Gamma_4(t).$$

We can keep track of the species concentration by dividing the number of molecules by the volume Ω. Letting $m = X_m/\Omega$, $P = X_P/\Omega$, and $\alpha_0 = \alpha X_D/\Omega$, we obtain the final expression

$$\frac{d}{dt}\begin{pmatrix} m \\ P \end{pmatrix} = \begin{pmatrix} -\delta & 0 \\ \kappa & -\gamma \end{pmatrix}\begin{pmatrix} m \\ P \end{pmatrix} + \begin{pmatrix} \alpha_0 \\ 0 \end{pmatrix} + \frac{1}{\sqrt{\Omega}}\begin{pmatrix} \left(\sqrt{\alpha_0 + \delta m}\right)\Gamma_m \\ \left(\sqrt{\kappa m + \gamma P}\right)\Gamma_P \end{pmatrix},$$

where Γ_m and Γ_P are independent Gaussian white noise processes (note that here we have used that if Γ_1 and Γ_2 are independent identical Gaussian white noise processes, then $\sqrt{a}\Gamma_1 + \sqrt{b}\Gamma_2 = \sqrt{a+b}\Gamma$ with Γ a Gaussian white noise process identical to Γ_i). $\qquad\qquad\qquad\qquad\qquad\qquad\qquad\qquad\qquad\qquad\qquad\qquad\nabla$

The Langevin equation formulation is particularly useful as it allows us to study the stochastic properties of the system by studying how the state responds to a (stochastic) input. Hence, a few of the tools available for studying input/output dynamic behavior can be employed (see Section 3.1, Section 3.2, and Section 4.3).

Fokker-Planck equations (FPE)

The chemical Langevin equation provides a stochastic ordinary differential equation that describes the evolution of the system state. A slightly different (but completely equivalent) representation of the dynamics is to model how the probability distribution $P(x, t)$ evolves in time. As in the case of the chemical Langevin equation, we will assume that the system state is continuous and write down a formula for the evolution of the density function $p(x, t)$. This formula is known as the Fokker-Planck equation (FPE) and is essentially an approximation to the chemical master equation.

Consider first the case of a random process in one dimension. We assume that the random process is in the same form as in the previous section:

$$\frac{dX(t)}{dt} = A(X(t)) + B(X(t))\Gamma(t). \tag{4.12}$$

The function $A(X)$ is called the *drift term* and $B(X)$ is the diffusion term. It can be shown that the probability density function for X, $p(x, t)$, satisfies the partial differential equation

$$\frac{\partial p}{\partial t}(x, t) = -\frac{\partial}{\partial x}(A(x, t)p(x, t)) + \frac{1}{2}\frac{\partial^2}{\partial x^2}(B^2(x, t)p(x, t)). \tag{4.13}$$

Note that here we have shifted to the probability density function since we are considering X to be a continuous state random process.

In the multivariate case, more care is required. Using the chemical Langevin equation (4.11), we define

$$D_{ij}(x,t) = \frac{1}{2} \sum_{k=1}^{M} B_{ik}(x,t) B_{jk}(x,t), \qquad i < j = 1, \ldots, M.$$

The Fokker-Planck equation now becomes

$$\frac{\partial p}{\partial t}(x,t) = -\sum_{i=1}^{M} \frac{\partial}{\partial x_i}(A_i(x,t)p(x,t)) + \sum_{i=1}^{M}\sum_{j=1}^{M} \frac{\partial^2}{\partial x_i \partial x_j}(D_{ij}(x,t)p(x,t)). \quad (4.14)$$

Note that the Fokker-Planck equation is very similar to the chemical master equation: both provide a description of how the probability distribution varies as a function of time. In the case of the Fokker-Planck equation, we regard the state as a continuous set of variables and we write a partial differential equation for how the probability density function evolves in time. In the case of the chemical master equation, we have a discrete state (microstates) and we write an ordinary differential equation for how the probability distribution (formally the probability mass function) evolves in time. Both formulations contain the same basic information, just using slightly different representations of the system and the probability of being in a given state.

Reaction rate equations (RRE)

As we already saw in Chapter 2, the reaction rate equations can be used to describe the dynamics of a chemical system in the case where there are a large number of molecules whose state can be approximated using just the concentrations of the molecules. We re-derive the results from Section 2.1 here, being more careful to point out what approximations are being made.

We start with the chemical Langevin equation (4.11), which has the form

$$\frac{dX_i(t)}{dt} = \sum_{j=1}^{M} \xi_{ji} a_j(X(t)) + \sum_{j=1}^{M} \xi_{ji} a_j^{1/2}(X(t)) \Gamma_j(t).$$

While we have not derived this expression in detail, we note that the first term simply says that the value of the random variable X_i fluctuates according to possible reaction vectors ξ_{ji} scaled by the probability that reaction j occurs in time dt.

We are now interested in how the mean of the concentration X_i evolves. Writing $\langle X_i \rangle$ for the mean (taken over many different samples of the random process), the dynamics of the species at each point in time are given by

$$\frac{d\langle X_i(t) \rangle}{dt} = \sum_{j=1}^{M} \xi_{ji} \langle a_j(X(t)) \rangle, \qquad (4.15)$$

where the second term in the Langevin equation drops out under the assumption that the Γ_j's are independent processes with zero mean. We see that the reaction rate equations follow by defining $x_i = \langle X_i \rangle / \Omega$ and *assuming* that $\langle a_j(X(t)) \rangle = a_j(\langle X(t) \rangle)$. This relationship is true when a_j is linear (e.g., in the case of a unimolecular reaction), but is an approximation otherwise.

The formal derivation of the reaction rate equations from the chemical master equation and the chemical Langevin equation requires a number of careful assumptions (see the original work of Gillespie [34] for a full derivation). In particular, it requires that the chemical system is well-stirred (no spatial structure), that the molecular counts are sufficiently high that we can approximate concentrations with real numbers, and that the time scales of interest are appropriately long so that multiple individual reactions can be appropriately averaged, and yet at the same time sufficiently short so that we can approximate the derivative through a finite different approximation. As we have noted previously, most biological systems have significant spatial structure (thus violating the well-stirred assumption), but models based on that assumption are still very useful in many settings. The larger molecular count assumption is more critical in using the reaction rate equation and one must take care when molecular counts are in the single digits, for example.

4.2 Simulation of stochastic systems

Suppose that we want to generate a collection of sample trajectories for a stochastic system whose evolution is described by the chemical master equation (4.7):

$$\frac{d}{dt} P(q,t) = \sum_i a_i(q - \xi_i) P(q - \xi_i, t) - \sum_i a_i(q) P(q, t),$$

where $P(q,t)$ is the probability of being in a microstate q at time t (starting from q_0 at time t_0) and $a_i(q)$ is the propensity function for a reaction i starting at a microstate q and ending at microstate $q + \xi_i$. Instead of simulating the distribution function $P(q,t)$, we wish to simulate a specific instance $q(t)$ starting from some initial condition $q_0(t_0)$. If we simulate many such instances of $q(t)$, their distribution at time t should match $P(q,t)$.

The stochastic simulation algorithm

The stochastic simulation algorithm is a Monte Carlo procedure for numerically generating time trajectories of the number of molecules of the various species present in the system in accordance with the chemical master equation.

To illustrate the basic ideas that we will use, consider first a simple birth process in which the microstate is given by an integer $q \in \{0, 1, 2, \dots\}$ and we assume that the propensity function is given by

$$a(q) \, dt = \lambda \, dt, \qquad \xi = +1.$$

Thus the probability of transition increases linearly with the time increment dt (so birth events occur at rate λ, on average). If we assume that the birth events are independent of each other, then it can be shown that the number of arrivals in time τ is Poisson distributed with parameter $\lambda\tau$:

$$P(q(t+\tau) - q(t) = \ell) = \frac{(\lambda\tau)^\ell}{\ell!} e^{-\lambda\tau},$$

where τ is the difference in time and ℓ is the difference in count q. In fact, this distribution is a joint distribution in time τ and count ℓ. Setting $\ell = 1$, it can be shown that the time to the next reaction, T, follows an exponential distribution and hence has density function

$$p_T(\tau) = \lambda e^{-\lambda\tau}.$$

The exponential distribution has expectation $1/\lambda$ and so we see that the average time between events is inversely proportional to the reaction rate λ.

Consider next a more general case in which we have a countable number of microstates $q \in \{0, 1, 2, \ldots\}$ and we let k_{ji} represent the transition probability between a microstate i and microstate j. The birth process is a special case given by $k_{i+1,i} = \lambda$ and all other $k_{ji} = 0$. The chemical master equation describes the joint probability that we are in state $q = i$ at a particular time t. We would like to know the probability that we transition to a new state $q = j$ at time $t + dt$. Given this probability, we can attempt to generate an instance of the variable $q(t)$ by first determining which reaction occurs and then when the reaction occurs.

Let $P(j, \tau) := P(j, t + \tau + d\tau \mid i, t + \tau)$ represent the probability that we transition from the state i to the state j in the time interval $[t + \tau, t + \tau + d\tau]$. For simplicity and ease of notation, we will take $t = 0$. Let $T := T_{j,i}$ be the time at which the reaction first occurs. We can write the probability that we transition to state j in the interval $[\tau, \tau + d\tau]$ as

$$P(j, \tau) = P(T > \tau) k_{ji} d\tau, \tag{4.16}$$

where $P(T > \tau)$ is the probability that no reaction occurs in the time interval $[0, \tau]$ and $k_{ji} d\tau$ is the probability that the reaction taking state i to state j occurs in the next $d\tau$ seconds (assumed to be independent events, giving the product of these probabilities).

To compute $P(T > \tau)$, define

$$\bar{k}_i = \sum_j k_{ji},$$

so that $(1 - \bar{k}_i) d\tau$ is the probability that no transition occurs from state i in the next $d\tau$ seconds. Then, the probability that no reaction occurs in the interval $[\tau, \tau + d\tau]$ can be written as

$$P(T > \tau + d\tau) = P(T > \tau)(1 - \bar{k}_i) \, d\tau. \tag{4.17}$$

It follows that

$$\frac{d}{d\tau}P(T > \tau) = \lim_{d\tau \to 0} \frac{P(T > \tau + d\tau) - P(T > \tau)}{d\tau} = -P(T > \tau)\,\bar{k}_i.$$

Solving this differential equation, we obtain

$$P(T > \tau) = e^{-\bar{k}_i\tau}, \tag{4.18}$$

so that the probability that no reaction occurs in time τ decreases exponentially with the amount of time that we wait, with rate given by the sum of all the reactions that can occur from state i.

We can now combine equation (4.18) with equation (4.16) to obtain

$$P(j, \tau) = P(j, \tau + d\tau \mid i, 0) = k_{ji}\, e^{-\bar{k}_i\tau}\, d\tau.$$

We see that this has the form of a density function in time and hence the probability that the next reaction is reaction j, independent of the time in which it occurs, is

$$P_{ji} = \int_0^\infty k_{ji} e^{-\bar{k}_i\tau}\, d\tau = \frac{k_{ji}}{\bar{k}_i}. \tag{4.19}$$

Thus, to choose the next reaction to occur from a state i, we choose between N possible reactions, with the probability of each reaction weighted by $k_{ji}/\bar{k}_i$.

To determine the time that the next reaction occurs, we sum over all possible reactions j to get the density function for the reaction time:

$$p_T(\tau) = \sum_j k_{ji} e^{-\bar{k}_i\tau} = \bar{k}_i e^{-\bar{k}_i\tau}.$$

This is the density function associated with an exponential distribution. To compute a time of reaction Δt that draws from this distribution, we note that the cumulative distribution function for T is given by

$$\int_0^{\Delta t} f_T(\tau)\, d\tau = \int_0^{\Delta t} \bar{k}_i e^{-\bar{k}_i\tau}\, d\tau = 1 - e^{-\bar{k}_i\Delta t}.$$

The cumulative distribution function is always in the range $[0, 1]$ and hence we can compute Δt by choosing a (uniformly distributed) random number r in $[0, 1]$ and then computing

$$\Delta t = \frac{1}{\bar{k}_i} \ln \frac{1}{1 - r}. \tag{4.20}$$

(This equation can be simplified somewhat by replacing $1 - r$ with r' and noting that r' can also be drawn from a uniform distribution on $[0, 1]$.)

Note that in the case of a birth process, this computation agrees with our earlier analysis. Namely, $\bar{k}_i = \lambda$ and hence the (only) reaction occurs according to an exponential distribution with parameter λ.

This set of calculations gives the following algorithm for computing an instance of the chemical master equation:

1. Choose an initial condition q at time $t = 0$.

2. Calculate the propensity functions $a_i(q)$ for each possible reaction i.

3. Choose the time for the reaction according to equation (4.20), where $r \in [0, 1]$ is chosen from a uniform distribution.

4. Use a weighted random number generator to identify which reaction will take place next, using the weights in equation (4.19).

5. Update q by implementing the reaction ξ and update the time t by Δt

6. If $T < T_{\text{stop}}$, go to step 2.

This method is sometimes called "Gillespie's direct method" [32, 33], but we shall refer to it here as the "stochastic simulation algorithm" (SSA). We note that the reaction number in step 4 can be computed by calculating a uniform random number on $[0, 1]$, scaling this by the total propensity $\sum_i a_i(q)$, and then finding the first reaction i such that $\sum_{j=0}^{i} a_j(q)$ is larger than this scaled random number.

4.3 Input/output linear stochastic systems

In many situations, we wish to know how noise propagates through a biomolecular system. For example, we may wish to understand how stochastic variations in RNA polymerase concentration affect gene expression. In order to analyze these cases, it is useful to make use of tools from stochastic control theory that allow analysis of noise propagation around a fixed operating point.

We begin with the chemical Langevin equation (4.11), which we can write as

$$\frac{dX(t)}{dt} = A(X(t)) + B(X(t))\Gamma(t).$$

The vector $X(t)$ consists of the individual random variables $X_i(t)$ representing the concentration of species S_i, the functions $A(X(t))$ and $B(X(t))$ are computed from the reaction vectors and propensity functions for the system, and Γ is a set of "white noise" processes. For the remainder of this chapter, we will assume that the function $A(X)$ is linear in X and that $B(X)$ is constant (by appropriately linearizing around the mean state, if needed). We will also rewrite Γ as W, to be more consistent with the literature of stochastic control systems.

Random processes

It will be useful in characterizing the properties of the vector $X(t)$ to treat it as a random process. We briefly review the basic definitions here, primarily to fix the terminology we will use in the rest of the section.

A *continuous-time random process* is a stochastic system characterized by the evolution of a random variable $X(t)$, $t \in [0, T]$. We are interested in understanding how the (random) state of the system is related at separate times, i.e., how the two random variables $X(t_1)$ and $X(t_2)$ are related. We characterize the state of a random process using a (joint) time-varying probability density function p:

$$\mathbb{P}(\{x_{i,l} \leq X_i(t) \leq x_{i,u}\}) = \int_{x_{1,l}}^{x_{1,u}} \cdots \int_{x_{n,l}}^{x_{n,u}} p_{X_1,\ldots,X_n}(x;t) dx_n \ldots dx_1.$$

Note that the state of a random process is not enough to determine the exact next state, but only the distribution of next states (otherwise it would be a deterministic process). We typically omit indexing of the individual states unless the meaning is not clear from context.

In general, the distributions used to describe a random process depend on the specific time or times that we evaluate the random variables. However, in some cases the relationship only depends on the difference in time and not the absolute times (similar to the notion of time invariance in deterministic systems, as described in Åström and Murray [1]). A process is *stationary* if the distribution is not changing and joint density functions only depend on the differences in times. More formally, $p(x, t + \tau) = p(x, t)$ for all τ, $p(x_i, x_j; t_1 + \tau, t_2 + \tau) = p(x_i, x_j; t_1, t_2)$, etc. In this case we can write $p(x_i, x_j; \tau)$ for the joint probability distribution. Stationary distributions roughly correspond to the steady state properties of a random process and we will often restrict our attention to this case.

$$\mathbb{E}(X(t)) = \begin{pmatrix} \mathbb{E}(X_1(t)) \\ \vdots \\ \mathbb{E}(X_n(t)) \end{pmatrix} =: \mu(t),$$

$$\mathbb{E}((X(t) - \mu(t))(X(t) - \mu(t))^T) =$$
$$\begin{pmatrix} \mathbb{E}((X_1(t) - \mu_1(t))(X_1(t) - \mu_1(t))) & \cdots & \mathbb{E}((X_1(t) - \mu_1(t))(X_n(t) - \mu_n(t))) \\ & \ddots & \vdots \\ & & \mathbb{E}((X_n(t) - \mu_n(t))(X_n(t) - \mu_n(t))) \end{pmatrix} =: \Sigma(t),$$

$$\mathbb{E}(X(t)X^T(s)) = \begin{pmatrix} \mathbb{E}(X_1(t)X_1(s)) & \cdots & \mathbb{E}(X_1(t)X_n(s)) \\ & \ddots & \vdots \\ & & \mathbb{E}(X_n(t)X_n(s)) \end{pmatrix} =: R(t, s).$$

Note that the random variables and their statistical properties are all indexed by the time t (and s). The matrix $R(t, s)$ is called the *correlation matrix* for $X(t) \in \mathbb{R}^n$. If $t = s$ then $R(t, t)$ describes how the elements of x are correlated at time t (with each other) and in the case that the processes have zero mean, $R(t, t) = \Sigma(t)$. The elements on the diagonal of $\Sigma(t)$ are the variances of the corresponding scalar variables. A random process is uncorrelated if $R(t, s) = 0$ for all $t \neq s$. This implies that $X(t)$ and $X(s)$ are uncorrelated random events and is equivalent to $p_{X,Y}(x, y) = p_X(x)p_Y(y)$.

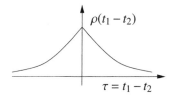

Figure 4.4: Correlation function for a first-order Markov process.

If a random process is stationary, then it can be shown that $R(t+\tau, s+\tau) = R(t,s)$ and it follows that the correlation matrix depends only on $t - s$. In this case we will often write $R(t, s) = R(s - t)$ or simply $R(\tau)$ where τ is the correlation time. The covariance matrix in this case is simply $R(0)$.

In the case where X is a scalar random process, the correlation matrix is also a scalar and we will write $r(\tau)$, which we refer to as the (scalar) correlation function. Furthermore, for stationary scalar random processes, the correlation function depends only on the absolute value of the correlation time, so $r(\tau) = r(-\tau) = r(|\tau|)$. This property also holds for the diagonal entries of the correlation matrix since $R_{ii}(s,t) = R_{ii}(t,s)$ from the definition.

Example 4.6 (Ornstein-Uhlenbeck process). Consider a scalar random process defined by a Gaussian probability density function with $\mu = 0$,

$$p(x,t) = \frac{1}{\sqrt{2\pi\sigma^2}}e^{-\frac{1}{2}\frac{x^2}{\sigma^2}},$$

and a correlation function given by

$$r(t_1,t_2) = \frac{Q}{2\omega_0}e^{-\omega_0|t_2-t_1|}.$$

The correlation function is illustrated in Figure 4.4. This process is known as an *Ornstein-Uhlenbeck process* and it is a stationary process.   ▽

Note on terminology. The terminology and notation for covariance and correlation varies between disciplines. The term covariance is often used to refer to both the relationship between different variables X and Y and the relationship between a single variable at different times, $X(t)$ and $X(s)$. The term "cross-covariance" is used to refer to the covariance between two random vectors X and Y, to distinguish this from the covariance of the elements of X with each other. The term "cross-correlation" is sometimes also used. Finally, the term "correlation coefficient" refers to the normalized correlation $\bar{r}(t, s) = \mathbb{E}(X(t)X(s))/\mathbb{E}(X(t)X(t))$.

We will also make use of a special type of random process referred to as "white noise." A *white noise process* $X(t)$ satisfies $\mathbb{E}(X(t)) = 0$ and $R(t, s) = W\delta(s - t)$, where $\delta(\tau)$ is the impulse function and W is called the *noise intensity*. White noise is an idealized process, similar to the impulse function or Heaviside (step) function in deterministic systems. In particular, we note that $r(0) = \mathbb{E}(X^2(t)) = \infty$, so the covariance is infinite and we never see this signal in practice. However, like the step and impulse functions, it is very useful for characterizing the response of a linear system, as described in the following proposition.

Linear stochastic systems with Gaussian noise

We now consider the problem of how to compute the response of a linear system to a random process. We assume we have a linear system described in state space as

$$\frac{dX}{dt} = AX + FW, \qquad Y = CX. \tag{4.21}$$

For simplicity, we take W and Y to be scalar random variables. Given an "input" W, which is itself a random process with mean $\mu(t)$, variance $\sigma^2(t)$ and correlation $r(t, t + \tau)$, what is the description of the random process Y?

Let W be a white noise process, with zero mean and noise intensity Q:

$$r(\tau) = Q\delta(\tau).$$

We can write the output of the system in terms of the convolution integral

$$Y(t) = \int_0^t h(t - \tau)W(\tau)\,d\tau,$$

where $h(t - \tau)$ is the impulse response for the system

$$h(t - \tau) = Ce^{A(t-\tau)}F.$$

We now compute the statistics of the output, starting with the mean:

$$\begin{aligned}
\mathbb{E}(Y(t)) &= \mathbb{E}\left(\int_0^t h(t - \eta)W(\eta)\,d\eta\right) \\
&= \int_0^t h(t - \eta)\mathbb{E}(W(\eta))\,d\eta = 0.
\end{aligned}$$

Note here that we have relied on the linearity of the convolution integral to pull the expectation inside the integral.

We can compute the covariance of the output by computing the correlation $r_Y(\tau)$ and setting $\sigma_Y^2 = r_Y(0)$. The correlation function for y is

$$\begin{aligned}
r_Y(t, s) = \mathbb{E}(Y(t)Y(s)) &= \mathbb{E}\left(\int_0^t h(t - \eta)W(\eta)\,d\eta \cdot \int_0^s h(s - \xi)W(\xi)\,d\xi\right) \\
&= \mathbb{E}\left(\int_0^t \int_0^s h(t - \eta)W(\eta)W(\xi)h(s - \xi)\,d\eta\,d\xi\right),
\end{aligned}$$

where we assume W is a scalar (otherwise $W(\xi)$ and $h(s-\xi)$ must be transposed).
Once again linearity allows us to exchange expectation with the integral and

$$
\begin{aligned}
r_Y(t,s) &= \int_0^t \int_0^s h(t-\eta)\mathbb{E}(W(\eta)W(\xi))h(s-\xi)\,d\eta d\xi \\
&= \int_0^t \int_0^s h(t-\eta)Q\delta(\eta-\xi)h(s-\xi)\,d\eta d\xi \\
&= \int_0^t h(t-\eta)Qh(s-\eta)\,d\eta.
\end{aligned}
$$

Now let $\tau = s - t$ and write

$$
\begin{aligned}
r_Y(\tau) = r_Y(t,t+\tau) &= \int_0^t h(t-\eta)Qh(t+\tau-\eta)\,d\eta \\
&= \int_0^t h(\xi)Qh(\xi+\tau)\,d\xi \qquad \text{(setting } \xi = t-\eta).
\end{aligned}
$$

Finally, we let $t \to \infty$ (steady state)

$$
\lim_{t\to\infty} r_Y(t,t+\tau) = \bar{r}_Y(\tau) = \int_0^\infty h(\xi)Qh(\xi+\tau)d\xi. \tag{4.22}
$$

If this integral exists, then we can compute the second-order statistics for the output
Y.

We can provide a more explicit formula for the correlation function r in terms of
the matrices A, F and C by expanding equation (4.22). We will consider the general
case where $W \in \mathbb{R}^p$ and $Y \in \mathbb{R}^q$ and use the correlation matrix $R(t,s)$ instead of the
correlation function $r(t,s)$. Define the *state transition matrix* $\Phi(t,t_0) = e^{A(t-t_0)}$ so
that the solution of system (4.21) is given by

$$
x(t) = \Phi(t,t_0)x(t_0) + \int_{t_0}^t \Phi(t,\lambda)FW(\lambda)d\lambda.
$$

Proposition 4.1 (Stochastic response to white noise). *Let $\mathbb{E}(X(t_0)X^T(t_0)) = P(t_0)$
and W be white noise with $\mathbb{E}(W(\lambda)W^T(\xi)) = R_W\delta(\lambda-\xi)$. Then the correlation matrix for X is given by*

$$
R_X(t,s) = P(t)\Phi^T(s,t)
$$

where $P(t)$ satisfies the linear matrix differential equation

$$
\dot{P}(t) = AP + PA^T + FR_WF, \qquad P(0) = P_0.
$$

The correlation matrix for the output Y can be computed using the fact that
$Y = CX$ and hence $R_Y = CR_XC^T$. We will often be interested in the steady state
properties of the output, which are given by the following proposition.

Proposition 4.2 (Steady state response to white noise). *For a time-invariant linear system driven by white noise, the correlation matrices for the state and output converge in steady state to*

$$R_X(\tau) = R_X(t, t+\tau) = Pe^{A^T\tau}, \qquad R_Y(\tau) = CR_X(\tau)C^T$$

where P satisfies the algebraic equation

$$AP + PA^T + FR_WF^T = 0 \qquad P > 0. \tag{4.23}$$

Equation (4.23) is called the *Lyapunov equation* and can be solved in MATLAB using the function `lyap`.

Example 4.7 (First-order system). Consider a scalar linear process

$$\dot{X} = -aX + W, \qquad Y = cX,$$

where W is a white, Gaussian random process with noise intensity σ^2. Using the results of Proposition 4.1, the correlation function for X is given by

$$R_X(t, t+\tau) = p(t)e^{-a\tau}$$

where $p(t) > 0$ satisfies

$$\frac{dp(t)}{dt} = -2ap(t) + \sigma^2.$$

We can solve explicitly for $p(t)$ since it is a (non-homogeneous) linear differential equation:

$$p(t) = e^{-2at}p(0) + (1 - e^{-2at})\frac{\sigma^2}{2a}.$$

Finally, making use of the fact that $Y = cX$ we have

$$r(t, t+\tau) = c^2(e^{-2at}p(0) + (1 - e^{-2at})\frac{\sigma^2}{2a})e^{-a\tau}.$$

In steady state, the correlation function for the output becomes

$$r(\tau) = \frac{c^2\sigma^2}{2a}e^{-a\tau}.$$

Note that the correlation function has the same form as the Ornstein-Uhlenbeck process in Example 4.6 (with $Q = c^2\sigma^2$). $\qquad\qquad\nabla$

Random processes in the frequency domain

As in the case of deterministic linear systems, we can analyze a stochastic linear system either in the state space or the frequency domain. The frequency domain approach provides a very rich set of tools for modeling and analysis of interconnected systems, relying on the frequency response and transfer functions to represent the flow of signals around the system.

Given a random process $X(t)$, we can look at the frequency content of the properties of the response. In particular, if we let $\rho(\tau)$ be the correlation function for a (scalar) random process, then we define the *power spectral density function* as the Fourier transform of ρ:

$$S(\omega) = \int_{-\infty}^{\infty} \rho(\tau) e^{-j\omega\tau} \, d\tau, \qquad \rho(\tau) = \frac{1}{2\pi} \int_{-\infty}^{\infty} S(\omega) e^{j\omega\tau} \, d\tau.$$

The power spectral density provides an indication of how quickly the values of a random process can change through the frequency content: if there is high frequency content in the power spectral density, the values of the random variable can change quickly in time.

Example 4.8 (Ornstein-Uhlenbeck process). To illustrate the use of these measures, consider the Ornstein-Uhlenbeck process whose correlation function we computed in Example 4.7:

$$\rho(\tau) = \frac{Q}{2\omega_0} e^{-\omega_0(\tau)}.$$

The power spectral density becomes

$$S(\omega) = \int_{-\infty}^{\infty} \frac{Q}{2\omega_0} e^{-\omega|\tau|} e^{-j\omega\tau} \, d\tau$$

$$= \int_{-\infty}^{0} \frac{Q}{2\omega_0} e^{(\omega-j\omega)\tau} \, d\tau + \int_{0}^{\infty} \frac{Q}{2\omega_0} e^{(-\omega-j\omega)\tau} \, d\tau = \frac{Q}{\omega^2 + \omega_0^2}.$$

We see that the power spectral density is similar to a transfer function and we can plot $S(\omega)$ as a function of ω in a manner similar to a Bode plot, as shown in Figure 4.5. Note that although $S(\omega)$ has a form similar to a transfer function, it is a real-valued function and is not defined for complex ω. ∇

Using the power spectral density, we can give a more intuitive definition of "white noise" as a zero-mean, random process with power spectral density $S(\omega) =$ constant for all ω. If $X(t) \in \mathbb{R}^n$ (a random vector), then $S(\omega) \in \mathbb{R}^{n \times n}$. We see that a random process is white if all frequencies are equally represented in its power spectral density; this spectral property is the reason for the terminology "white."

Given a linear system

$$\dot{X} = AX + FW, \qquad Y = CX,$$

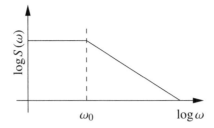

Figure 4.5: Spectral power density for a first-order Markov process.

with W given by white noise, we can compute the spectral density function corresponding to the output Y. Let $H(s) = C(sI - A)^{-1}B$ be the transfer function from W to Y. We start by computing the Fourier transform of the steady state correlation function (4.22):

$$
\begin{aligned}
S_Y(\omega) &= \int_{-\infty}^{\infty} \left[\int_0^{\infty} h(\xi) Q h(\xi + \tau) d\xi \right] e^{-j\omega\tau} \, d\tau \\
&= \int_0^{\infty} h(\xi) Q \left[\int_{-\infty}^{\infty} h(\xi + \tau) e^{-j\omega\tau} \, d\tau \right] d\xi \\
&= \int_0^{\infty} h(\xi) Q \left[\int_0^{\infty} h(\lambda) e^{-j\omega(\lambda - \xi)} \, d\lambda \right] d\xi \\
&= \int_0^{\infty} h(\xi) e^{j\omega\xi} \, d\xi \cdot Q H(j\omega) = H(-j\omega) Q H(j\omega).
\end{aligned}
$$

This is then the (steady state) response of a linear system to white noise.

As with transfer functions, one of the advantages of computations in the frequency domain is that the composition of two linear systems can be represented by multiplication. In the case of the power spectral density, if we pass white noise through a system with transfer function $H_1(s)$ followed by transfer function $H_2(s)$, the resulting power spectral density of the output is given by

$$
S_Y(\omega) = H_1(-j\omega) H_2(-j\omega) Q_u H_2(j\omega) H_1(j\omega).
$$

Exercises

4.1 Consider a standard model of transcription and translation with probabilistic creation and degradation of discrete mRNA and protein molecules. The *propensity functions* for each reaction are as follows:

- Probability of transcribing 1 mRNA molecule: $0.2dt$

- Probability of degrading 1 mRNA molecule: $0.5dt$ and is proportional to the number of mRNA molecules.

- Probability of translating 1 protein: $5dt$ and is proportional to the number of mRNA molecules.

- Probability of degrading 1 protein molecule: $0.5dt$ and is proportional to the number of protein molecules.

In each case, dt will be the time step chosen for your simulation, which we take as $dt = 0.05$ sec.

(i) Simulate the stochastic system above until time $T = 100$. Plot the resulting number of mRNA and protein over time.

(ii) Now assume that the proteins are degraded much more slowly than mRNA and the propensity function of protein degradation is now $0.05dt$. To maintain similar protein levels, the translation probability is now $0.5dt$ (and still proportional to the number of mRNA molecules). Simulate this system as above. What difference do you see in protein level? Comment on the effect of protein degradation rates on noise.

4.2 Compare a simple model of negative autoregulation to one without autoregulation:

$$\frac{dX}{dt} = \beta_0 - \gamma X$$

and

$$\frac{dX}{dt} = \frac{\beta}{1 + X/K} - \gamma X.$$

(i) Assume that the basal transcription rates β and β_0 vary between cells, following a Gaussian distribution with $\sigma/\mu = 0.1$. Simulate time courses of both models for 100 different "cells" using the following parameters: $\beta = 2, \beta_0 = 1, \gamma = 1, K = 1$. Plot the nonregulated and autoregulated systems in two separate plots. Comment on the variation you see in the time courses.

(ii) Calculate the deterministic steady state for both models above. How does variation in the basal transcription rate β or β_0 enter into the steady state? Relate it to what you see in part (i).

4.3 Consider a simple model for gene expression with reactions

$$\phi \xrightarrow{\alpha} m, \qquad m \xrightarrow{\kappa} m + P, \qquad m \xrightarrow{\delta} \phi, \qquad P \xrightarrow{\gamma} \emptyset.$$

Let $\alpha = 1/2, \kappa = 20\log(2)/120, \delta = \log(2)/120$ and $\gamma = \log(2)/600$, and answer the following questions:

(i) Use the stochastic simulation algorithm (SSA) to obtain realizations of the stochastic process of gene expression and numerically compare with the deterministic ODE solution. Explore how the realizations become close to or apart from the ODE solution when the volume is changed. Determine the stationary probability distribution for the protein (you can do this numerically).

(ii) Now consider the additional binding reaction of protein P with downstream DNA binding sites D:

$$P + D \underset{d}{\overset{a}{\rightleftharpoons}} C.$$

Note that the system is no longer linear due to the presence of a bimolecular reaction. Use the SSA algorithm to obtain sample realizations and numerically compute the probability distribution of the protein. Compare it to what you obtained in part (i). Explore how this probability distribution and the one of C change as the rate constants a and d become larger with respect to $\gamma, \alpha, \kappa, \delta$. Do you think we can use a QSS approximation similar to what we have done for ODE models?

(iii) Determine the Langevin equation for the system in part (ii) and obtain sample realizations. Explore numerically how good this approximation is when the volume decreases/increases.

4.4 Consider the bimolecular reaction

$$A + B \underset{d}{\overset{a}{\rightleftharpoons}} C,$$

in which A and B are in total amounts A_{tot} and B_{tot}, respectively. Compare the steady state value of C obtained from the deterministic model to the mean value of C obtained from the stochastic model as the volume is changed in the stochastic model. What do you observe? You can perform this investigation through numerical simulation.

4.5 Consider the simple birth and death process:

$$Z \underset{k_1 G}{\overset{k_2 G}{\rightleftharpoons}} \emptyset,$$

in which G is a "gain." Assume that the reactions are catalyzed by enzymes and that the gain G can be tuned by changing the amounts of these enzymes. A deterministic ODE model for this system incorporating disturbances due to environmental perturbations is given by

$$\frac{dZ}{dt} = k_1 G - k_2 GZ + d(t).$$

Determine the Langevin equation for this birth and death process and compare its form to the deterministic one. Also, determine the frequency response of Z to noise for both the deterministic model and for the Langevin model. Does increasing the gain G have the same effect in both models? Explain.

4.6 Consider a second-order system with dynamics

$$\frac{d}{dt}\begin{pmatrix} X_1 \\ X_2 \end{pmatrix} = \begin{pmatrix} -a & 0 \\ 0 & -b \end{pmatrix}\begin{pmatrix} X_1 \\ X_2 \end{pmatrix} + \begin{pmatrix} 1 \\ 1 \end{pmatrix} w, \qquad Y = \begin{pmatrix} 1 & 1 \end{pmatrix}\begin{pmatrix} X_1 \\ X_2 \end{pmatrix}$$

that is forced by Gaussian white noise w with zero mean and variance σ^2. Assume $a, b > 0$.

(i) Compute the correlation function $\rho(\tau)$ for the output of the system. Your answer should be an explicit formula in terms of a, b and σ.

(ii) Assuming that the input transients have died out, compute the mean and variance of the output.

Chapter 5
Biological Circuit Components

In this chapter, we describe some simple circuit components that have been constructed in *E. coli* cells using the technology of synthetic biology and then consider a more complicated circuit that already appears in natural systems to implement adaptation. We will analyze the behavior of these circuits employing mainly the tools from Chapter 3 and some of the tools from Chapter 4. The basic knowledge of Chapter 2 will be assumed.

5.1 Introduction to biological circuit design

In Chapter 2 we introduced a number of core processes and models for those processes, including gene expression, transcriptional regulation, post-translational regulation such as covalent modification of proteins, allosteric regulation of enzymes, and activity regulation of transcription factors through inducers. These core processes provide a rich set of functional building blocks, which can be combined together to create circuits with prescribed functionalities.

For example, if we want to create an inverter, a device that returns high output when the input is low and vice versa, we can use a gene regulated by a transcriptional repressor. If we want to create a signal amplifier, we can employ a cascade of covalent modification cycles. Specifically, if we want the amplifier to be linear, we should tune the amounts of protein substrates to be smaller than the Michaelis-Menten constants. Alternatively, we could employ a phosphotransfer system, which provides a fairly linear input/output relationship for an extended range of the input stimulation. If instead we are looking for an almost digital response, we could employ a covalent modification cycle with high amounts of substrates compared to the Michaelis-Menten constants. Furthermore, if we are looking for a fast input/output response, phosphorylation cycles are better candidates than transcriptional systems.

In this chapter and the next we illustrate how one can build circuits with prescribed functionality using some of the building blocks of Chapter 2 and the design techniques illustrated in Chapter 3. We will focus on two types of circuits: gene circuits and signal transduction circuits. In some cases, we will illustrate designs that incorporate both.

A gene circuit is usually depicted by a set of nodes, each representing a gene, connected by unidirectional edges, representing a transcriptional activation or a re-

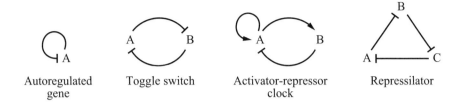

Figure 5.1: Early gene circuits that have been fabricated in bacteria *E. coli*: the negatively autoregulated gene [10], the toggle switch [30], the activator-repressor clock [6], and the repressilator [26].

pression. If gene z represses the expression of gene x, the interaction is represented by $Z \dashv X$. If instead gene z activates the expression of gene x, the interaction is represented by $Z \rightarrow X$. Inducers will often appear as additional nodes, which activate or inhibit a specific edge. Early examples of such circuits include an autoregulated circuit [10], a toggle switch obtained by connecting two inverters in a ring fashion [30], an activator-repressor system that can display toggle switch or clock behavior [6], and a loop oscillator called the repressilator obtained by connecting three inverters in a ring topology [26] (Figure 5.1).

Basic synthetic biology technology

Simple synthetic gene circuits can be constituted from a set of (connected) transcriptional components, which are made up by the DNA base pair sequences that compose the desired promoters, ribosome binding sites, gene coding region, and terminators. We can choose these components from a library of basic parts, which are classified based on biochemical properties such as affinity (of promoter, operator, or ribosome binding sites), strength (of a promoter), and efficiency (of a terminator).

The desired sequence of parts is usually assembled on plasmids, which are circular pieces of DNA, separate from the host cell chromosome, with their own origin of replication. These plasmids are then inserted, through a process called transformation in bacteria and transfection in yeast, in the host cell. Once in the host cell, they express the proteins they code for by using the transcription and translation machinery of the cell. There are three main types of plasmids: low copy number (5-10 copies), medium copy number (15-20 copies), and high copy number (up to hundreds). The copy number reflects the average number of copies of the plasmid inside the host cell. The higher the copy number, the more efficient the plasmid is at replicating itself. The exact number of plasmids in each cell fluctuates stochastically and cannot be exactly controlled.

In order to measure the amounts of proteins of interest, we make use of *reporter genes*. A reporter gene codes for a protein that fluoresces in a specific color (red, blue, green, or yellow, for example) when it is exposed to light of the appropriate wavelength. For instance, green fluorescent protein (GFP) is a protein with the property that it fluoresces in green when exposed to UV light. It is produced by the jellyfish *Aequoria victoria* and its gene has been isolated so that it can be used as a reporter. Other fluorescent proteins, such as yellow fluorescent protein (YFP) and red fluorescent protein (RFP), are genetic variations of GFP.

A reporter gene is usually inserted downstream of the gene expressing the protein whose concentration we want to measure. In this case, both genes are under the control of the same promoter and are transcribed into a single mRNA molecule. The mRNA is then translated to protein and the two proteins will be fused together. This technique provides a direct way to measure the concentration of the protein of interest but can affect the functionality of this protein because some of its regulatory sites may be occluded by the fluorescent protein. Another viable technique is one in which the reporter gene is placed under the control of the same promoter that is also controlling the expression of the protein of interest. In this case, the production rates of the reporter and of the protein of interest are the same and, as a consequence, the respective concentrations should mirror each other. The reporter thus provides an indirect measurement of the concentration of the protein of interest.

Just as fluorescent proteins can be used as a readout of a circuit, inducers function as external inputs that can be used to probe the system. Two commonly used negative inducers are IPTG and aTc, as explained in Section 2.3, while two common positive inducers are arabinose and AHL. Arabinose activates the transcriptional activator AraC, which activates the pBAD promoter. Similarly, AHL is a signaling molecule that activates the LuxR transcription factor, which activates the pLux promoter.

Protein dynamics can usually be altered by the addition of a degradation tag at the end of the corresponding coding region. A degradation tag is a sequence of base pairs that adds an amino acid sequence to the functional protein that is recognized by proteases. Proteases then bind to the protein, degrading it into a non-functional molecule. As a consequence, the half-life of the protein decreases, resulting in an increased decay rate. Degradation tags are often employed to obtain a faster response of the protein concentration to input stimulation and to prevent protein accumulation.

5.2 Negative autoregulation

In this section, we analyze the negatively autoregulated gene of Figure 5.1 and focus on analyzing how the presence of the negative feedback affects the dynamics and the noise properties of the system. This system was introduced in Example 2.2.

Let A be a transcription factor repressing its own production. Assuming that the mRNA dynamics are at the quasi-steady state, the ODE model describing the negatively autoregulated system is given by

$$\frac{dA}{dt} = \frac{\beta}{1 + (A/K)^n} - \gamma A. \tag{5.1}$$

We seek to compare the behavior of this autoregulated system, which we also refer to as the closed loop system, to the behavior of the unregulated one:

$$\frac{dA}{dt} = \beta_0 - \gamma A,$$

in which β_0 is the unrepressed production rate. We refer to this second system as the open loop system.

Dynamic effects of negative autoregulation

As we showed via simulation in Example 2.2, negative autoregulation speeds up the response to perturbations. Hence, the time the system takes to reach its equilibrium decreases with negative feedback. In this section, we illustrate how this result can be analytically demonstrated by employing linearization. Specifically, we linearize the system about its equilibrium point and calculate the time response resulting from initializing the linearized system at an initial condition close to the equilibrium point.

Let $A_e = \beta_0/\gamma$ be the equilibrium of the unregulated system and let $z = A - A_e$ denote the perturbation with respect to such an equilibrium. The dynamics of z are given by

$$\frac{dz}{dt} = -\gamma z.$$

Given a small initial perturbation z_0, the response of z is given by the exponential

$$z(t) = z_0 e^{-\gamma t}.$$

The "half-life" of the signal $z(t)$ is the time $z(t)$ takes to reach half of z_0 and we denote it by t_{half}. This is a common measure for the speed of response of a system to an initial perturbation. Simple mathematical calculation shows that $t_{\text{half}} = \ln(2)/\gamma$. Note that the half-life does not depend on the production rate β_0 and only depends on the protein decay rate constant γ.

Now let A_e be the steady state of the negatively autoregulated system (5.1). Assuming that the perturbation z with respect to the equilibrium is small enough, we can employ linearization to describe the dynamics of z. These dynamics are given by

$$\frac{dz}{dt} = -\bar{\gamma} z,$$

where

$$\bar{\gamma} = \gamma + \beta \frac{nA_e^{n-1}/K^n}{(1 + (A_e/K)^n)^2}.$$

In this case, we have that $t_{\text{half}} = \ln(2)/\bar{\gamma}$.

Since $\bar{\gamma} > \gamma$ (for any positive value of A_e), we have that the dynamic response to a perturbation is faster in the system with negative autoregulation. This confirms the simulation findings of Example 2.2.

Noise filtering

We next investigate the effect of the negative autoregulation on the noisiness of the system. In order to do this, we employ the Langevin modeling framework and determine the frequency response to the intrinsic noise on the various reactions. In particular, in the analysis that follows we treat Langevin equations as regular ordinary differential equations with inputs, allowing us to apply the tools described in Chapter 3.

We perform two different studies. In the first one, we assume that the decay rate of the protein is much slower than that of the mRNA. As a consequence, the mRNA concentration can be well approximated by its quasi-steady state value and we focus on the dynamics of the protein only. In the second study, we investigate the consequence of having the mRNA and protein decay rates in the same range so that the quasi-steady state assumption cannot be made. This can be the case, for example, when degradation tags are added to the protein to make its decay rate larger. In either case, we study both the open loop system and the closed loop system (the system with negative autoregulation) and compare the corresponding frequency responses to noise.

Assuming that mRNA is at its quasi-steady state

In this case, the reactions for the open loop system are given by

$$\text{R1}: \quad \text{p} \xrightarrow{\beta_0} \text{A} + \text{p}, \qquad \text{R2}: \quad \text{A} \xrightarrow{\gamma} \emptyset,$$

in which β_0 is the constitutive production rate, p is the DNA promoter, and γ is the decay rate of the protein. Since the concentration of DNA promoter p is not changed by these reactions, it is a constant, which we call p_{tot}.

Employing the Langevin equation (4.11) of Section 4.1 and letting n_A denote the real-valued number of molecules of A and n_p the real-valued number of molecules of p, we obtain

$$\frac{dn_A}{dt} = \beta_0 n_p - \gamma n_A + \sqrt{\beta_0 n_p} \, \Gamma_1 - \sqrt{\gamma n_A} \, \Gamma_2,$$

in which Γ_1 and Γ_2 depend on the noise on the production reaction and on the decay reaction, respectively. By letting $A = n_A/\Omega$ denote the concentration of A and $p = n_p/\Omega = p_{\text{tot}}$ denote the concentration of p, we have that

$$\frac{dA}{dt} = \beta_0 p_{\text{tot}} - \gamma A + \frac{1}{\sqrt{\Omega}} (\sqrt{\beta_0 p_{\text{tot}}} \, \Gamma_1 - \sqrt{\gamma A} \, \Gamma_2). \tag{5.2}$$

This is a linear system and therefore we can calculate the frequency response to any of the two inputs Γ_1 and Γ_2. In particular, the frequency response to input Γ_1 has magnitude given by

$$M^o(\omega) = \frac{\sqrt{\beta_0 p_{\text{tot}}/\Omega}}{\sqrt{\omega^2 + \gamma^2}}. \tag{5.3}$$

We now consider the autoregulated system. The reactions are given by

$$\text{R1}: \quad p \xrightarrow{\beta} A + p, \qquad \text{R2}: \quad A \xrightarrow{\gamma} \emptyset,$$

$$\text{R3}: \quad A + p \xrightarrow{a} C, \qquad \text{R4}: \quad C \xrightarrow{d} A + p.$$

Defining $p_{\text{tot}} = p + C$ and employing the Langevin equation (4.11) of Section 4.1, we obtain

$$\frac{dp}{dt} = -aAp + d(p_{\text{tot}} - p) + \frac{1}{\sqrt{\Omega}}(-\sqrt{aAp}\,\Gamma_3 + \sqrt{d(p_{\text{tot}} - p)}\,\Gamma_4),$$

$$\frac{dA}{dt} = \beta p - \gamma A - aAp + d(p_{\text{tot}} - p) + \frac{1}{\sqrt{\Omega}}(\sqrt{\beta p}\,\Gamma_1 - \sqrt{\gamma A}\,\Gamma_2 - \sqrt{aAp}\,\Gamma_3$$

$$+ \sqrt{d(p_{\text{tot}} - p)}\,\Gamma_4),$$

in which Γ_3 and Γ_4 are the noises corresponding to the association and dissociation reactions, respectively. Letting $K_d = d/a$,

$$N_1 = \frac{1}{\sqrt{\Omega}}(-\sqrt{Ap/K_d}\,\Gamma_3 + \sqrt{(p_{\text{tot}} - p)}\,\Gamma_4), \qquad N_2 = \frac{1}{\sqrt{\Omega}}(\sqrt{\beta p}\,\Gamma_1 - \sqrt{\gamma A}\,\Gamma_2),$$

we can rewrite the above system in the following form:

$$\frac{dp}{dt} = -aAp + d(p_{\text{tot}} - p) + \sqrt{d}N_1(t),$$

$$\frac{dA}{dt} = \beta p - \gamma A - aAp + d(p_{\text{tot}} - p) + N_2(t) + \sqrt{d}N_1(t).$$

Since $d \gg \gamma, \beta$, this system displays two time scales. Letting $\epsilon := \gamma/d$ and defining $y := A - p$, the system can be rewritten in standard singular perturbation form (3.24):

$$\epsilon\frac{dp}{dt} = -\gamma Ap/K_d + \gamma(p_{\text{tot}} - p) + \sqrt{\epsilon}\,\sqrt{\gamma}N_1(t),$$

$$\frac{dy}{dt} = \beta p - \gamma(y + p) + N_2(t).$$

By setting $\epsilon = 0$, we obtain the quasi-steady state value $p = p_{\text{tot}}/(1 + A/K_d)$. Writing $\dot{A} = \dot{y} + \dot{p}$, using the chain rule for $\dot{p}$, and assuming that p_{tot}/K_d is sufficiently small, we obtain the reduced system describing the dynamics of A as

$$\frac{dA}{dt} = \beta\frac{p_{\text{tot}}}{1 + A/K_d} - \gamma A + \frac{1}{\sqrt{\Omega}}\left(\sqrt{\beta\frac{p_{\text{tot}}}{1 + A/K_d}}\,\Gamma_1 - \sqrt{\gamma A}\,\Gamma_2\right) =: f(A, \Gamma_1, \Gamma_2). \tag{5.4}$$

The equilibrium point for this system corresponding to the mean values $\Gamma_1 = 0$ and $\Gamma_2 = 0$ of the inputs is given by

$$A_e = \frac{1}{2}\left(\sqrt{K_d^2 + 4\beta p_{tot} K_d/\gamma} - K_d\right).$$

The linearization of the system about this equilibrium point is given by

$$\left.\frac{\partial f}{\partial A}\right|_{A_e,\Gamma_1=0,\Gamma_2=0} = -\beta\frac{p_{tot}/K_d}{(1 + A_e/K_d)^2} - \gamma =: -\bar{\gamma},$$

$$b_1 = \left.\frac{\partial f}{\partial \Gamma_1}\right|_{A_e,\Gamma_1=0,\Gamma_2=0} = \frac{1}{\sqrt{\Omega}}\sqrt{\frac{\beta p_{tot}}{1 + A_e/K_d}}, \quad b_2 = \left.\frac{\partial f}{\partial \Gamma_2}\right|_{A_e,\Gamma_1=0,\Gamma_2=0} = -\frac{1}{\sqrt{\Omega}}\sqrt{\gamma A_e}.$$

Hence, the frequency response to Γ_1 has magnitude given by

$$M^c(\omega) = \frac{b_1}{\sqrt{\omega^2 + \bar{\gamma}^2}}. \tag{5.5}$$

In order to make a fair comparison between this response and that of the open loop system, we need the equilibrium points of both systems to be the same. In order to guarantee this, we set β such that

$$\frac{\beta}{1 + A_e/K_d} = \beta_0.$$

This can be attained, for example, by properly adjusting the strength of the promoter and of the ribosome binding site. As a consequence, we have that $b_1 = \sqrt{\beta_0 p_{tot}/\Omega}$. Since we also have that $\bar{\gamma} > \gamma$, comparing expressions (5.3) and (5.5) it follows that $M^c(\omega) < M^o(\omega)$ for all ω. That is, the magnitude of the frequency response of the closed loop system is smaller than that of the open loop system at all frequencies. Hence, negative autoregulation attenuates noise at all frequencies. The two frequency responses are plotted in Figure 5.2a. A similar result could be obtained for the frequency response with respect to the input Γ_2 (see Exercise 5.1).

mRNA decay close to protein decay

In the case in which mRNA and protein decay rates are comparable, we need to model both the processes of transcription and translation. Letting m_A denote the mRNA of A, the reactions describing the open loop system modify to

$$\text{R1:} \quad m_A \xrightarrow{\kappa} m_A + A, \qquad \text{R2:} \quad A \xrightarrow{\gamma} \emptyset,$$

$$\text{R5:} \quad p \xrightarrow{\alpha_0} m_A + p, \qquad \text{R6:} \quad m_A \xrightarrow{\delta} \emptyset,$$

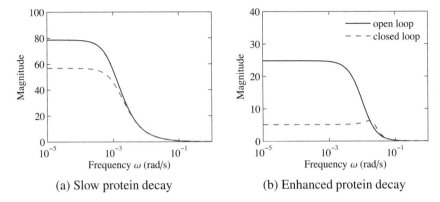

(a) Slow protein decay   (b) Enhanced protein decay

Figure 5.2: Effect of negative autoregulation on noise propagation. (a) Magnitude of the frequency response to noise $\Gamma_1(t)$ for both open loop and closed loop systems for the model in which mRNA is assumed at its quasi-steady state. The parameters are $p_{tot} = 10$ nM, $K_d = 10$ nM, $\beta = 0.001$ min^{-1}, $\gamma = 0.001$ min^{-1}, and $\beta_0 = 0.00092$ min^{-1}, and we assume unit volume Ω. (b) Frequency response to noise $\Gamma_6(t)$ for both open loop and closed loop for the model in which mRNA decay is close to protein decay. The parameters are $p_{tot} = 10$ nM, $K_d = 10$ nM, $\alpha = 0.001$ min^{-1}, $\beta = 0.01$ min^{-1}, $\delta = 0.01$ min^{-1}, $\gamma = 0.01$ min^{-1}, and $\alpha_0 = 0.0618$ min^{-1}.

while those describing the closed loop system become

$$R1: \quad m_A \xrightarrow{\kappa} m_A + A, \qquad R2: \quad A \xrightarrow{\gamma} \emptyset,$$

$$R3: \quad A + p \xrightarrow{a} C, \qquad R4: \quad C \xrightarrow{d} A + p,$$

$$R5: \quad p \xrightarrow{\alpha} m_A + p, \qquad R6: \quad m_A \xrightarrow{\delta} \emptyset.$$

Defining $p_{tot} = p + C$, employing the Langevin equation, and applying singular perturbation as performed before, we obtain the dynamics of the system as

$$\frac{dm_A}{dt} = F(A) - \delta m_A + \frac{1}{\sqrt{\Omega}}(\sqrt{F(A)}\,\Gamma_5 - \sqrt{\delta m_A}\,\Gamma_6),$$

$$\frac{dA}{dt} = \kappa m_A - \gamma A + \frac{1}{\sqrt{\Omega}}(\sqrt{\kappa m_A}\,\Gamma_1 - \sqrt{\gamma A}\,\Gamma_2),$$

in which Γ_5 and Γ_6 model the noise on the production reaction and decay reaction of mRNA, respectively. For the open loop system we have $F(A) = \alpha_0 p_{tot}$, while for the closed loop system we have the Hill function

$$F(A) = \frac{\alpha p_{tot}}{1 + A/K_d}.$$

The equilibrium point for the open loop system is given by

$$m_e^o = \frac{\alpha_0 p_{tot}}{\delta}, \qquad A_e^o = \frac{\kappa \alpha_0 p_{tot}}{\delta \gamma}.$$

Considering Γ_6 as the input of interest, the linearization of the system at this equilibrium is given by

$$A^o = \begin{pmatrix} -\delta & 0 \\ \kappa & -\gamma \end{pmatrix}, \qquad B^o = \begin{pmatrix} \sqrt{\delta m_e^o / \Omega} \\ 0 \end{pmatrix}.$$

Letting $K = \kappa/(\gamma K_d)$, the equilibrium for the closed loop system is given by

$$A_e^c = \frac{\kappa m_e^c}{\gamma}, \qquad m_e^c = \frac{1}{2}\left(-1/K + \sqrt{(1/K)^2 + 4\alpha p_{tot}/(K\delta)}\right).$$

The linearization of the closed loop system at this equilibrium point is given by

$$A^c = \begin{pmatrix} -\delta & -g \\ \kappa & -\gamma \end{pmatrix}, \qquad B^c = \begin{pmatrix} \sqrt{\delta m_e^c / \Omega} \\ 0 \end{pmatrix}, \qquad (5.6)$$

in which $g = (\alpha p_{tot}/K_d)/(1 + A_e^c/K_d)^2$ represents the contribution of the negative autoregulation. The larger the value of g—obtained, for example, by making K_d smaller (see Exercise 5.2)—the stronger the negative autoregulation.

In order to make a fair comparison between the open loop and closed loop system, we again set the equilibrium points to be the same. To do this, we choose α such that $\alpha/(1 + A_e^c/K_d) = \alpha_0$, which can be done by suitably changing the strengths of the promoter and ribosome binding site.

The open loop and closed loop transfer functions are thus given by

$$G_{A\Gamma_6}^o(s) = \frac{\kappa\sqrt{\delta m_e/\Omega}}{(s+\delta)(s+\gamma)}, \qquad G_{A\Gamma_6}^c(s) = \frac{\kappa\sqrt{\delta m_e/\Omega}}{s^2 + s(\delta+\gamma) + \delta\gamma + \kappa g}.$$

From these expressions, it follows that the open loop transfer function has two real poles, while the closed loop transfer function can have complex conjugate poles when g is sufficiently large. As a consequence, noise Γ_6 can be amplified at sufficiently high frequencies. Figure 5.2b shows the magnitude $M(\omega)$ of the corresponding frequency responses for both the open loop and the closed loop systems.

It follows that the presence of negative autoregulation attenuates noise with respect to the open loop system at low frequency, but it can amplify noise at higher frequency. This is a very well-studied phenomenon known as the "waterbed effect," according to which negative feedback decreases the effect of disturbances at low frequency, but it can amplify it at higher frequency. This effect is not found in first-order models, as demonstrated by the derivations performed when mRNA is at the quasi-steady state. This illustrates the spectral shift of the frequency response to intrinsic noise towards the high frequency, as also experimentally reported [7].

5.3 The toggle switch

The toggle switch is composed of two genes that mutually repress each other, as shown in the diagram of Figure 5.1. We start by describing a simple model with no

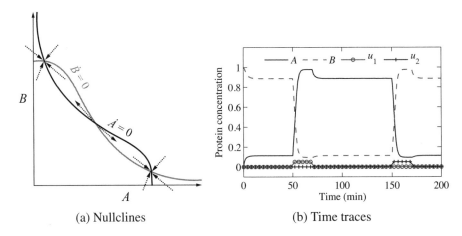

(a) Nullclines   (b) Time traces

Figure 5.3: Genetic toggle switch. (a) Nullclines for the toggle switch. By analyzing the
direction of the vector field in the proximity of the equilibria, one can deduce their stability
as described in Section 3.1. (b) Time traces for $A(t)$ and $B(t)$ when inducer concentrations
$u_1(t)$ and $u_2(t)$ are changed. The plots show a scaled version of these signals, whose absolute values are $u_1 = u_2 = 1$ in the indicated intervals of time. In the simulation, we have
$n = 2$, $K_{d,1} = K_{d,2} = 1$ nM, $K = \sqrt{0.1}$ nM, $\beta = 1$ hrs^{-1}, and $\gamma = 1$ hrs^{-1}.

inducers. By assuming that the mRNA dynamics are at the quasi-steady state, we
obtain a two-dimensional differential equation model given by

$$\frac{dA}{dt} = \frac{\beta}{1 + (B/K)^n} - \gamma A, \qquad \frac{dB}{dt} = \frac{\beta}{1 + (A/K)^n} - \gamma B,$$

in which we have assumed for simplicity that the parameters of the repression
functions are the same for A and B.

Since the system is two-dimensional, both the number and stability of equilibria
can be analyzed by performing nullcline analysis (see Section 3.1). Specifically, by
setting $dA/dt = 0$ and $dB/dt = 0$ and letting $n \geq 2$, we obtain the nullclines shown
in Figure 5.3a. The nullclines intersect at three points, which determine the equilibrium points of this system. The stability of these equilibria can be determined
by the following graphical reasoning.

The nullclines partition the plane into six regions. By determining the sign of
dA/dt and dB/dt in each of these six regions, we can determine the direction in
which the vector field is pointing in each of these regions. From these directions,
we can deduce that the equilibrium occurring for intermediate values of A and B
(at which $A = B$) is unstable while the other two are stable (see the arrows in Figure
5.3a). Hence, the toggle switch is a bistable system.

The system trajectories converge to one equilibrium or the other depending on
whether the initial condition is in the region of attraction of the first or the second
equilibrium. The 45-degree line divides the plane into the two regions of attraction
of the stable equilibrium points. Once the system's trajectory has converged to

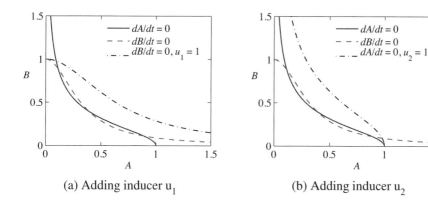

(a) Adding inducer u_1 (b) Adding inducer u_2

Figure 5.4: Genetic toggle switch with inducers. (a) Nullclines for the toggle switch (solid and dashed lines) and how the nullcline $\dot{B} = 0$ changes when inducer u_1 is added (dash-dotted line). (b) Nullclines for the toggle switch (solid line) and how the nullcline $\dot{A} = 0$ changes when inducer u_2 is added (dotted line). Parameter values are as in Figure 5.3.

one of the two equilibrium points, it cannot switch to the other unless an external (transient) stimulation is applied.

In the genetic toggle switch developed by Gardner et al. [30], external stimulations were added in the form of negative inducers for A and B. Specifically, let u_1 be the negative inducer for A and u_2 be the negative inducer for B. Then, as we have seen in Section 2.3, the expressions of the Hill functions need to be modified to replace A by $A(1/(1 + u_1/K_{d,1}))$ and B by $B(1/(1 + u_2/K_{d,2}))$, in which $K_{d,1}$ and $K_{d,2}$ are the dissociation constants of u_1 with A and of u_2 with B, respectively. Hence, the system dynamics become

$$\frac{dA}{dt} = \frac{\beta}{1 + (B/K_B(u_2))^n} - \gamma A, \qquad \frac{dB}{dt} = \frac{\beta}{1 + (A/K_A(u_1))^n} - \gamma B,$$

in which we have let $K_A(u_1) = K(1 + u_1/K_{d,1})$ and $K_B(u_2) = K(1 + u_2/K_{d,2})$ denote the effective K values of the Hill functions. We show in Figure 5.3b time traces for $A(t)$ and $B(t)$ when the inducer concentrations are changed. The system starts from initial conditions in which B is high and A is low without inducers. Then, at time 50 the system is presented with a short pulse in u_2, which causes A to rise since it prevents B to repress A. As A rises, B is repressed and hence B decreases to the low value and the system state switches to the other stable steady state. The system remains in this steady state also after the inducer u_2 is removed until another transient stimulus is presented at time 150. At this time, there is a pulse in u_1, which inhibits the ability of A to repress B and, as a consequence, B rises, thus repressing A, and the system returns to its original steady state.

Note that the effect of the inducers in this model is that of temporarily changing the shape of the nullclines by increasing the values of K_A and K_B. Specifically, high values of u_1 with $u_2 = 0$ will lead to increased values of K_A, which will shift the point of half-maximal value of the Hill function $\beta/(1 + (A/K_A)^n)$ to the right. As a consequence, the nullclines will intersect at one point only, in which the value of B is high and the value of A is low (Figure 5.4a). The opposite will occur when u_2 is high and $u_1 = 0$, leading to only one intersection point in which B is low and A is high (Figure 5.4b).

5.4   The repressilator

Elowitz and Leibler constructed an oscillatory genetic circuit consisting of three repressors arranged in a ring fashion and called it the "repressilator" [26] (Figure 5.1). The repressilator exhibits sinusoidal, limit cycle oscillations in periods of hours, slower than the cell-division time. Therefore, the state of the oscillator is transmitted between generations from mother to daughter cells.

A dynamical model of the repressilator can be obtained by composing three transcriptional modules in a loop fashion. The dynamics can be written as

$$\frac{dm_A}{dt} = F_1(C) - \delta m_A, \qquad \frac{dm_B}{dt} = F_2(A) - \delta m_B, \qquad \frac{dm_C}{dt} = F_3(B) - \delta m_C,$$
$$\frac{dA}{dt} = \kappa m_A - \gamma A, \qquad \frac{dB}{dt} = \kappa m_B - \gamma B, \qquad \frac{dC}{dt} = \kappa m_C - \gamma C, \tag{5.7}$$

where we take

$$F_1(P) = F_2(P) = F_3(P) = F(P) = \frac{\alpha}{1 + (P/K)^n},$$

and assume initially that the parameters are the same for all the three repressor modules. The structure of system (5.7) belongs to the class of cyclic feedback systems that we have studied in Section 3.3. In particular, the Mallet-Paret and Smith Theorem 3.5 and Hastings et al. Theorem 3.4 can be applied to infer that if the system has a unique equilibrium point and this equilibrium is unstable, then the system admits a periodic solution. Therefore, to apply these results, we determine the number of equilibria and their stability.

The equilibria of the system can be found by setting the time derivatives to zero. Letting $\beta = (\kappa/\delta)$, we obtain

$$A_{eq} = \frac{\beta F_1(C_{eq})}{\gamma}, \qquad B_{eq} = \frac{\beta F_2(A_{eq})}{\gamma}, \qquad C_{eq} = \frac{\beta F_3(B_{eq})}{\gamma},$$

which combined together yield

$$A_{eq} = \frac{\beta}{\gamma} F_1 \left(\frac{\beta}{\gamma} F_3 \left(\frac{\beta}{\gamma} F_2(A_{eq}) \right) \right) =: g(A_{eq}).$$

The solution to this equation determines the set of equilibria of the system. The number of equilibria is given by the number of crossings of the two functions

$h_1(A) = g(A)$ and $h_2(A) = A$. Since h_2 is strictly monotonically increasing, we obtain a unique equilibrium if h_1 is monotonically decreasing. This is the case when $g'(A) = dg(A)/dA < 0$, otherwise there could be multiple equilibrium points. Since we have that

$$\text{sign}(g'(A)) = \prod_{i=1}^{3} \text{sign}(F_i'(A)),$$

it follows that if $\Pi_{i=1}^{3} \text{sign}(F_i'(A)) < 0$ the system has a unique equilibrium. We call the product $\Pi_{i=1}^{3} \text{sign}(F_i'(A))$ the *loop sign*.

It follows that any cyclic feedback system with negative loop sign will have a unique equilibrium. In the present case, system (5.7) is such that $F_i' < 0$, so that the loop sign is negative and there is a unique equilibrium. We next study the stability of this equilibrium by studying the linearization of the system.

Letting P denote the equilibrium value of the protein concentrations for A, B, and C, the Jacobian matrix of the system is given by

$$J = \begin{pmatrix} -\delta & 0 & 0 & 0 & 0 & F_1'(P) \\ \kappa & -\gamma & 0 & 0 & 0 & 0 \\ 0 & F_2'(P) & -\delta & 0 & 0 & 0 \\ 0 & 0 & \kappa & -\gamma & 0 & 0 \\ 0 & 0 & 0 & F_3'(P) & -\delta & 0 \\ 0 & 0 & 0 & 0 & \kappa & -\gamma \end{pmatrix},$$

whose characteristic polynomial is given by

$$\det(sI - J) = (s+\gamma)^3(s+\delta)^3 - \kappa^3 \prod_{i=1}^{3} F_i'(P). \tag{5.8}$$

The roots of this characteristic polynomial are given by

$$(s+\gamma)(s+\delta) = r,$$

in which $r \in \{\kappa F'(P), -(\kappa F'(P)/2)(1 - i\sqrt{3}), -(\kappa F'(P)/2)(1 + i\sqrt{3})\}$ and $i = \sqrt{-1}$ represents the imaginary unit. In order to invoke Hastings et al. Theorem 3.4 to infer the existence of a periodic orbit, it is sufficient that one of the roots of the characteristic polynomial has positive real part. This is the case if

$$\kappa|F'(P)| > 2\gamma\delta, \qquad |F'(P)| = \alpha \frac{n(P^{n-1}/K^n)}{(1 + (P/K)^n)^2},$$

in which P is the equilibrium value satisfying the equilibrium condition

$$P = \frac{\beta}{\gamma} \frac{\alpha}{1 + (P/K)^n}.$$

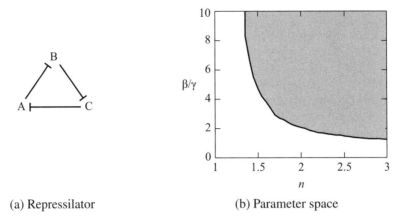

(a) Repressilator   (b) Parameter space

Figure 5.5: Parameter space for the repressilator. (a) Repressilator diagram. (b) Space of parameters that give rise to oscillations. Here, we have set $K = 1$ for simplicity.

One can plot the pair of values $(n, \beta/\gamma)$ for which the above two conditions are satisfied. This leads to the plot of Figure 5.5b. When n increases, the existence of an unstable equilibrium point is guaranteed for larger ranges of β/γ. Of course, this "behavioral" robustness does not guarantee that other important features of the oscillator, such as the period, are not changed when parameters vary.

A similar result for the existence of a periodic solution can be obtained when two of the Hill functions are monotonically increasing and only one is monotonically decreasing:

$$F_1(P) = \frac{\alpha}{1 + (P/K)^n}, \qquad F_2(P) = \frac{\alpha(P/K)^n}{1 + (P/K)^n}, \qquad F_3(P) = \frac{\alpha(P/K)^n}{1 + (P/K)^n}.$$

That is, two interactions are activations and one is a repression. We refer to this as the "non-symmetric" design. Since the loop sign is still negative, there is only one equilibrium point. We can thus obtain the condition for oscillations again by establishing conditions on the parameters that guarantee that at least one root of the characteristic polynomial (5.8) has positive real part, that is,

$$\kappa(|F_1'(P_3)F_2'(P_1)F_3'(P_2)|)^{(1/3)} > 2\gamma\delta, \tag{5.9}$$

in which P_1, P_2, P_3 are the equilibrium values of A, B, and C, respectively. These equilibrium values satisfy:

$$P_2 = \frac{\beta}{\gamma} \frac{(P_1/K)^n}{1 + (P_1/K)^n}, \qquad P_3 = \frac{\beta}{\gamma} \frac{(P_2/K)^n}{1 + (P_2/K)^n}, \qquad P_1(1 + (P_3/K)^n) = \frac{\beta}{\gamma}.$$

Using these expressions numerically and checking for each combination of the parameters $(n, \beta/\gamma)$ whether (5.9) is satisfied, we can plot the combinations of n and β/γ values that lead to an unstable equilibrium. This is shown in Figure 5.6b.

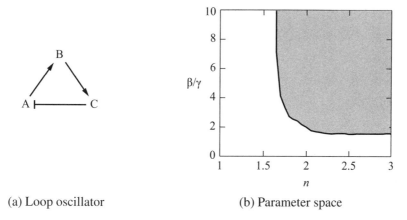

(a) Loop oscillator   (b) Parameter space

Figure 5.6: Parameter space for a loop oscillator. (a) Oscillator diagram. (b) Space of parameters that give rise to oscillations. As the value of n is increased, the range of the other parameter for which a periodic cycle exists becomes larger. Here, we have set $K = 1$.

From this figure, we can deduce that the qualitative shape of the parameter space that leads to a limit cycle is the same in the repressilator and in the non-symmetric design. One can conclude that it is then possible to design the circuit such that the parameters land in the filled region of the plots.

In practice, values of the Hill coefficient n between one and two can be obtained by employing repressors that have cooperativity higher than or equal to two. There are plenty of such repressors, including those originally used in the repressilator design [26]. However, values of n greater than two may be hard to reach in practice. To overcome this problem, one can include more elements in the loop. In fact, it is possible to show that the value of n sufficient for obtaining an unstable equilibrium decreases when the number of elements in the loop is increased (see Exercise 5.6). Figure 5.7a shows a simulation of the repressilator.

In addition to determining the space of parameters that lead to periodic trajectories, it is also relevant to determine the parameters to which the system behavior is the most sensitive. To address this question, we can use the parameter sensitivity analysis tools of Section 3.2. In this case, we model the repressilator Hill functions adding the basal expression rate as it was originally done in [26]:

$$F_1(P) = F_2(P) = F_3(P) = \frac{\alpha}{1 + (P/K)^n} + \alpha_0.$$

Letting $x = (m_A, A, m_B, B, m_C, C)$ and $\theta = (\alpha_0, \delta, \kappa, \gamma, \alpha, K)$, we can compute the sensitivity $S_{x,\theta}$ along the limit cycle corresponding to nominal parameter vector θ_0 as illustrated in Section 3.2:

$$\frac{dS_{x,\theta}}{dt} = M(t, \theta_0)S_{x,\theta} + N(t, \theta_0),$$

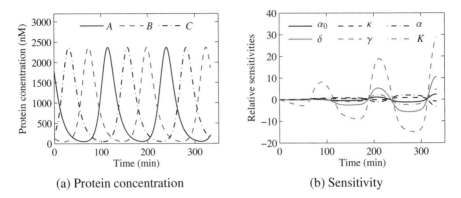

(a) Protein concentration   (b) Sensitivity

Figure 5.7: Repressilator parameter sensitivity analysis. (a) Protein concentrations as functions of time. (b) Sensitivity plots. The most important parameters are the protein and mRNA decay rates γ and δ. Parameter values used in the simulations are $\alpha = 800$ nM/s, $\alpha_0 = 5 \times 10^{-4}$ nM/s, $\delta = 5.78 \times 10^{-3}$ s^{-1}, $\gamma = 1.16 \times 10^{-3}$ s^{-1}, $\kappa = 0.116$ s^{-1}, $n = 2$, and $K = 1600$ nM.

where $M(t,\theta_0)$ and $N(t,\theta_0)$ are both periodic in time. If the dynamics of $S_{x,\theta}$ are stable then the resulting solutions will be periodic, showing how the dynamics around the limit cycle depend on the parameter values. The results are shown in Figure 5.7, where we plot the steady state sensitivity of A as a function of time. We see, for example, that the limit cycle depends strongly on the protein degradation and dilution rate δ, indicating that changes in this value can lead to (relatively) large variations in the magnitude of the limit cycle.

5.5 Activator-repressor clock

Consider the activator-repressor clock diagram shown in Figure 5.1. The activator A takes two inputs: the activator A itself and the repressor B. The repressor B has the activator A as the only input. Let m_A and m_B represent the mRNA of the activator and of the repressor, respectively. Then, we consider the following four-dimensional model describing the rate of change of the species concentrations:

$$\frac{dm_A}{dt} = F_1(A,B) - \delta_A m_A, \qquad \frac{dm_B}{dt} = F_2(A) - \delta_B m_B,$$
$$\frac{dA}{dt} = \kappa_A m_A - \gamma_A A, \qquad \frac{dB}{dt} = \kappa_B m_B - \gamma_B B, \qquad (5.10)$$

in which the functions F_1 and F_2 are Hill functions and given by

$$F_1(A,B) = \frac{\alpha_A(A/K_A)^n + \alpha_{A0}}{1 + (A/K_A)^n + (B/K_B)^m}, \quad F_2(A) = \frac{\alpha_B(A/K_A)^n + \alpha_{B0}}{1 + (A/K_A)^n}.$$

The Hill function F_1 can be obtained through a combinatorial promoter, where there are sites both for an activator and for a repressor. The Hill function F_2 has the

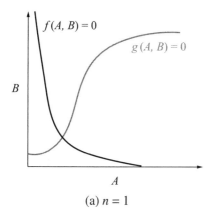
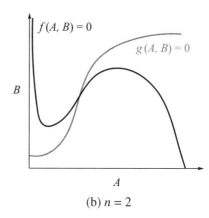

(a) $n = 1$ (b) $n = 2$

Figure 5.8: Nullclines for the two-dimensional system (5.11). Graph (a) shows the only possible configuration of the nullclines when $n = 1$. Graph (b) shows a possible configuration of the nullclines when $n = 2$. In this configuration, there is a unique equilibrium, which can be unstable.

form considered for an activator when transcription can still occur at a basal level even in the absence of an activator (see Section 2.3).

We first assume the mRNA dynamics to be at the quasi-steady state so that we can perform two-dimensional analysis and invoke the Poincaré-Bendixson theorem (Section 3.3). Then, we analyze the four-dimensional system and perform a bifurcation study.

Two-dimensional analysis

We let $f_1(A, B) := (\kappa_A/\delta_A)F_1(A, B)$ and $f_2(A) := (\kappa_B/\delta_B)F_2(A)$. For simplicity, we also define $f(A, B) := -\gamma_A A + f_1(A, B)$ and $g(A, B) := -\gamma_B B + f_2(A)$ so that the two-dimensional system is given by

$$\frac{dA}{dt} = f(A, B), \qquad \frac{dB}{dt} = g(A, B). \tag{5.11}$$

To simplify notation, we set $K_A = K_B = 1$ and take $m = 1$, without loss of generality as similar results can be obtained when $m > 1$ (see Exercise 5.7).

We first study whether the system admits a periodic solution for $n = 1$. To do so, we analyze the nullclines to determine the number and location of steady states. Let $\bar{\alpha}_A = \alpha_A(\kappa_A/\delta_A)$, $\bar{\alpha}_B = \alpha_B(\kappa_B/\delta_B)$, $\bar{\alpha}_{A0} = \alpha_{A0}(\kappa_A/\delta_A)$, and $\bar{\alpha}_{B0} = \alpha_{B0}(\kappa_B/\delta_B)$. Then, $g(A, B) = 0$ leads to

$$B = \frac{\bar{\alpha}_B A + \bar{\alpha}_{B0}}{(1 + A)\gamma_B},$$

which is an increasing function of A. Setting $f(A, B) = 0$, we obtain that

$$B = \frac{\bar{\alpha}_A A + \bar{\alpha}_{A0} - \gamma_A A(1 + A)}{\gamma_A A},$$

which is a monotonically decreasing function of A. These nullclines are displayed in Figure 5.8a.

We see that we have only one equilibrium point. To determine the stability of the equilibrium, we calculate the linearization of the system at such an equilibrium. This is given by the Jacobian matrix

$$
J = \begin{pmatrix} \dfrac{\partial f}{\partial A} & \dfrac{\partial f}{\partial B} \\[2ex] \dfrac{\partial g}{\partial A} & \dfrac{\partial g}{\partial B} \end{pmatrix}.
$$

In order for the equilibrium to be unstable and not a saddle, it is necessary and sufficient that $\mathrm{tr}(J) > 0$ and $\det(J) > 0$. Graphical inspection of the nullclines at the equilibrium (see Figure 5.8a) shows that

$$
\left. \frac{dB}{dA} \right|_{f(A,B)=0} < 0.
$$

By the implicit function theorem (Section 3.5), we further have that

$$
\left. \frac{dB}{dA} \right|_{f(A,B)=0} = -\frac{\partial f/\partial A}{\partial f/\partial B},
$$

so that $\partial f/\partial A < 0$ because $\partial f/\partial B < 0$. As a consequence, we have that $\mathrm{tr}(J) < 0$ and hence the equilibrium point is either stable or a saddle.

To determine the sign of $\det(J)$, we further inspect the nullclines and find that

$$
\left. \frac{dB}{dA} \right|_{g(A,B)=0} > \left. \frac{dB}{dA} \right|_{f(A,B)=0}.
$$

Again using the implicit function theorem we have that

$$
\left. \frac{dB}{dA} \right|_{g(A,B)=0} = -\frac{\partial g/\partial A}{\partial g/\partial B},
$$

so that $\det(J) > 0$. Hence, the ω-limit set (Section 3.3) of any point in the plane is necessarily not part of a periodic orbit. It follows that to guarantee that any initial condition converges to a periodic orbit, we need to require that $n > 1$.

We now study the case $n = 2$. In this case, the nullcline $f(A, B) = 0$ changes and can have the shape shown in Figure 5.8b. In the case in which, as in the figure, there is only one equilibrium point and the nullclines both have positive slope at the intersection (equivalent to $\det(J) > 0$), the equilibrium is unstable and not a saddle if $\mathrm{tr}(J) > 0$. This is the case when

$$
\frac{\gamma_B}{\partial f_1/\partial A - \gamma_A} < 1.
$$

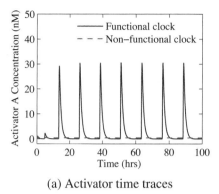

(a) Activator time traces

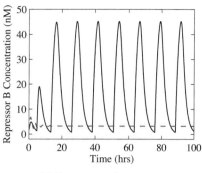

(b) Repressor time traces

Figure 5.9: Effect of the trace of the Jacobian on the stability of the equilibrium. The above plots illustrate the trajectories of system (5.11) for both a functional ($\mathrm{tr}(J) > 0$) and a non-functional ($\mathrm{tr}(J) < 0$) clock. The parameters in the simulation are $\delta_A = 1 = \delta_B = 1$ hrs^{-1}, $\alpha_A = 250$ nM/hrs, $\alpha_B = 30$ nM/hrs, $\alpha_{A0} = .04$ nM/hrs, $\alpha_{B0} = .004$ nM/hrs, $\gamma_A = 1$ hrs^{-1}, $\kappa_A = \kappa_B = 1$ hrs^{-1}, $K_A = K_B = 1$ nM, $n = 2$ and $m = 4$. In the functional clock, $\gamma_B = 0.5$ hrs^{-1}, whereas in the non-functional clock, $\gamma_B = 1.5$ hrs^{-1}.

This condition reveals the crucial design requirement for the functioning of the clock. Specifically, the repressor B time scale must be sufficiently slower than the activator A time scale. This point is illustrated in the simulations of Figure 5.9, in which we see that if γ_B is too large, the trace becomes negative and oscillations disappear.

Four-dimensional analysis

In order to deepen our understanding of the role of time scale separation between activator and repressor dynamics, we perform a time scale analysis employing the bifurcation tools described in Section 3.4. To this end, we consider the following four-dimensional model describing the rate of change of the species concentrations:

$$\frac{dm_A}{dt} = F_1(A, B) - (\delta_A/\epsilon)\, m_A, \qquad \frac{dm_B}{dt} = F_2(A) - (\delta_B/\epsilon)\, m_B,$$
$$\frac{dA}{dt} = \nu((\kappa_A/\epsilon)\, m_A - \gamma_A A), \qquad \frac{dB}{dt} = (\kappa_B/\epsilon)\, m_B - \gamma_B B. \tag{5.12}$$

This system is the same as system (5.10), where we have explicitly introduced two parameters ν and ϵ that model time scale differences, as follows. The parameter ν determines the relative time scale between the activator and the repressor dynamics. As ν increases, the activator dynamics become faster compared to the repressor dynamics. The parameter ϵ determines the relative time scale between the protein and mRNA dynamics. As ϵ becomes smaller, the mRNA dynamics become faster compared to protein dynamics and model (5.12) becomes close to the two-dimensional model (5.11), in which the mRNA dynamics are considered at

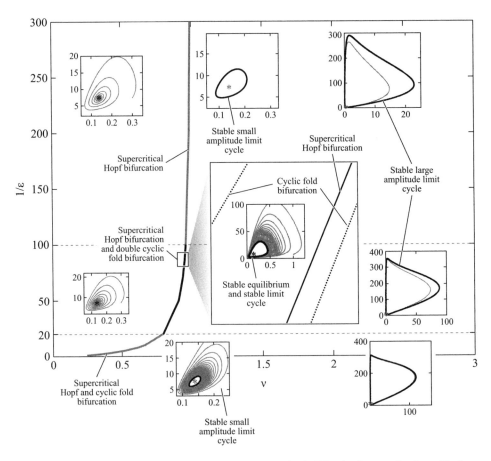

Figure 5.10: Design chart for the activator-repressor clock. We obtain sustained oscillations past the Hopf bifurcation point for values of ν sufficiently large independent of the difference of time scales between the protein and the mRNA dynamics. We also notice that there are values of ν for which a stable equilibrium point and a stable limit cycle coexist and values of ν for which two stable limit cycles coexist. The interval of ν values for which two stable limit cycles coexist is too small to be able to numerically set ν in such an interval. Thus, this interval is not practically relevant. The values of ν for which a stable equilibrium and a stable periodic orbit coexist are instead relevant. Figure adapted from [21].

the quasi-steady state. Thus, ϵ is a singular perturbation parameter. In particular, equations (5.12) can be taken to standard singular perturbation form by considering the change of variables $\bar{m}_A = m_A/\epsilon$ and $\bar{m}_B = m_B/\epsilon$. The details on singular perturbation can be found in Section 3.5.

The values of ϵ and of ν do not affect the number of equilibria of the system but they do determine the stability of the equilibrium points. We thus perform bifurcation analysis with ϵ and ν as the two bifurcation parameters. The bifurcation analysis results are summarized by Figure 5.10. In terms of the ϵ and ν parameters, it is thus possible to design the system as follows: if the activator dynamics are

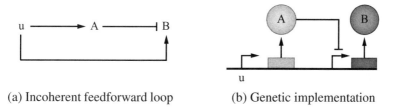

(a) Incoherent feedforward loop   (b) Genetic implementation

Figure 5.11: The incoherent feedforward loop (a) with a possible implementation (b). The circuit is integrated on a DNA plasmid denoted u. Protein A is under the control of a constitutive promoter in the DNA plasmid u, while B is repressed by A. Protein B, in turn, is also expressed by a gene in the plasmid u. Hence B is also "activated" by u.

sufficiently sped up with respect to the repressor dynamics, the system undergoes a Hopf bifurcation (Hopf bifurcation was introduced in Section 3.3) and a stable periodic orbit arises. The chart illustrates that for intermediate values of $1/\epsilon$, more complicated dynamic behaviors can arise in which a stable equilibrium coexists with a stable limit cycle. This situation corresponds to the *hard excitation* condition [60] and occurs for realistic values of the separation of time scales between protein and mRNA dynamics. Therefore, this simple oscillator motif described by a four-dimensional model can capture interesting dynamic behaviors, including features that lead to the long-term suppression of a rhythm by external inputs.

From a fabrication point of view, the activator dynamics can be sped up by adding suitable degradation tags to the activator protein. Similarly, the repressor dynamics can be slowed down by adding repressor DNA binding sites (see Chapter 6 and the effects of retroactivity on dynamic behavior).

5.6   An incoherent feedforward loop (IFFL)

In Section 3.2, we described various mechanisms to obtain robustness to external perturbations. In particular, one such mechanism is provided by incoherent feedforward loops. Here, we describe an implementation that was proposed for making the equilibrium values of protein expression robust to perturbations in DNA plasmid copy number [14]. In this implementation, the input u is the amount of DNA plasmid coding for both the intermediate regulator A and the output protein B. The intermediate regulator A represses the expression of the output protein B through transcriptional repression (Figure 5.11). The expectation is that the equilibrium value of B is independent of the concentration u of the plasmid. That is, the concentration of B should adapt to the copy number of its own plasmid.

In order to analyze whether the adaptation property holds, we write the differential equation model describing the system, assuming that the mRNA dynamics

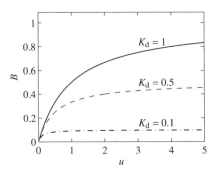

Figure 5.12: Behavior of the equilibrium value of B as a function of the input u. Concentration is in μM.

are at the quasi-steady state. This model is given by

$$\frac{dA}{dt} = k_0 u - \gamma A, \qquad \frac{dB}{dt} = \frac{k_1 u}{1 + (A/K_d)} - \gamma B, \qquad (5.13)$$

in which k_0 is the constitutive rate at which A is expressed and K_d is the dissociation constant of the binding of A with the promoter. This implementation has been called the sniffer in Section 3.2. The equilibrium of the system is obtained by setting the time derivatives to zero and gives

$$A = \frac{k_0}{\gamma} u, \qquad B = \frac{k_1 u}{\gamma + k_0 u / K_d}.$$

From this expression, we can see that as K_d decreases, the denominator of the right-hand side expression tends to $k_0 u / K_d$ resulting in the equilibrium value $B = k_1 K_d / k_0$, which does not depend on the input u. Hence, in this case, adaptation would be reached. This is the case if the affinity of A to its operator sites is extremely high, resulting also in a strong repression and hence a lower value of B. In practice, however, the value of K_d is nonzero, hence the adaptation is not perfect. We show in Figure 5.12 the equilibrium value of B as a function of the input u for different values of K_d. As expected, lower values of K_d lead to weaker dependence of B on the u variable.

In this analysis, we have not modeled the cooperativity of the binding of protein A to the promoter. We leave as an exercise to show that the adaptation behavior persists in the case cooperativity is included (see Exercise 5.8).

For engineering a system with prescribed behavior, one has to be able to change the physical features so as to change the values of the parameters of the model. This is often possible. For example, the binding affinity ($1/K_d$ in the Hill function) of a transcription factor to its site on the promoter can be weakened by single or multiple base-pair substitutions. The protein decay rate can be increased by adding degradation tags at the end of the gene expressing protein B. Promoters that can

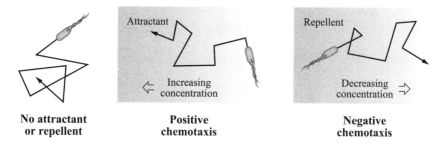

No attractant   **Positive**   **Negative**
or repellent   **chemotaxis**   **chemotaxis**

Figure 5.13: Examples of chemotaxis. In the absence of attractant or repellent, the bacterium follows a random walk. In the presence of an attractant, the random walk is biased in the direction in which the concentration of attractant increases (positive chemotaxis), while in the presence of a repellent the random walk is biased in the direction in which the concentration of the repellent decreases (negative chemotaxis). Figure adapted from Phillips, Kondev and Theriot [78].

accept multiple transcription factors (combinatorial promoters) can be realized by combining the operator sites of several simple promoters [46]. Finally, the overall protein production rate can be tuned by controlling a number of different system properties, including promoter's and the ribosome binding site's strength.

5.7   Bacterial chemotaxis

Chemotaxis refers to the process by which microorganisms move in response to chemical stimuli. Examples of chemotaxis include the ability of organisms to move in the direction of nutrients or move away from toxins in the environment. Chemotaxis is called *positive chemotaxis* if the motion is in the direction of the stimulus and *negative chemotaxis* if the motion is away from the stimulant, as shown in Figure 5.13. Many chemotaxis mechanisms are stochastic in nature, with biased random motions causing the average behavior to be either positive, negative or neutral (in the absence of stimuli).

In this section we look in some detail at bacterial chemotaxis, which *E. coli* use to move in the direction of increasing nutrients. The material in this section is based primarily on the work of Barkai and Leibler [9] and Rao, Kirby and Arkin [83].

Control system overview

The chemotaxis system in *E. coli* consists of a sensing system that detects the presence of nutrients, an actuation system that propels the organism in its environment, and control circuitry that determines how the cell should move in the presence of chemicals that stimulate the sensing system.

The actuation system in *E. coli* consists of a set of flagella that can be spun using a flagellar motor embedded in the outer membrane of the cell, as shown in

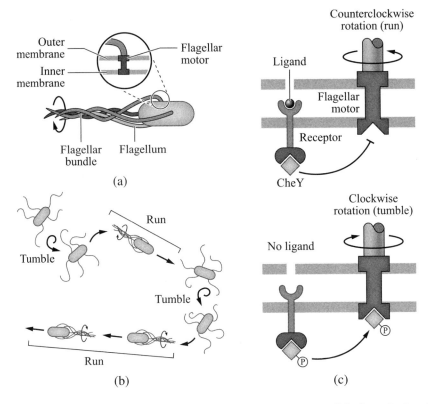

Figure 5.14: Bacterial chemotaxis. (a) Flagellar motors are responsible for spinning flagella. (b) When flagella spin in the clockwise direction, the organism tumbles, while when they spin in the counterclockwise direction, the organism runs. (c) The direction in which the flagella spin is determined by whether the CheY protein is phosphorylated. Figures adapted from Phillips, Kondev and Theriot [78].

Figure 5.14a. When the flagella all spin in the counterclockwise direction, the individual flagella form a bundle and cause the organism to move roughly in a straight line. This behavior is called a "run" motion. Alternatively, if the flagella spin in the clockwise direction, the individual flagella do not form a bundle and the organism "tumbles," causing it to rotate (Figure 5.14b). The selection of the motor direction is controlled by the protein CheY: if phosphorylated CheY binds to the motor complex, the motor spins clockwise (tumble), otherwise it spins counterclockwise (run) (Figure 5.14c).

A question that we address here is how the bacterium moves in the direction in which the attractant concentration increases. Because of the small size of the organism, it is not possible for a bacterium to sense gradients across its length. Hence, a more sophisticated strategy is used, in which the temporal gradient, as opposed to the spatial gradient, guides the organism motion through suitable combination of run and tumble motions. To sense temporal gradients, *E. coli* compares the cur-

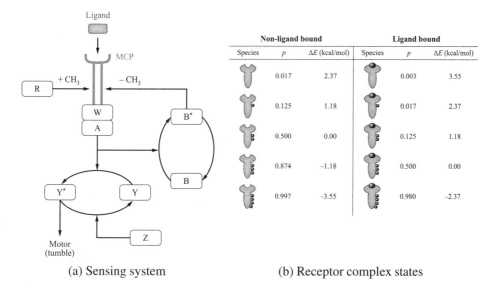

	Non-ligand bound			Ligand bound	
Species	p	ΔE (kcal/mol)	Species	p	ΔE (kcal/mol)
	0.017	2.37		0.003	3.55
	0.125	1.18		0.017	2.37
	0.500	0.00		0.125	1.18
	0.874	−1.18		0.500	0.00
	0.997	−3.55		0.980	−2.37

(a) Sensing system   (b) Receptor complex states

Figure 5.15: Control system for chemotaxis. (a) The sensing system is implemented by the receptor complex, which detects the presence of a ligand in the cell's outer environment. The computation part is implemented by a combined phosphorylation/methylation process, which realizes a form of integral feedback. The actuation component is realized by the CheY phosphorylated protein controlling directly the direction in which the motor spins. Figure from Rao et al. [83] (Figure 1A). (b) Receptor complex states. The probability of a given state being in an active configuration is given by p. ΔE represents the difference in energy levels from a reference state. Figure adapted from [72].

rent concentration of attractant to the past concentration of attractant and if the concentration increases, then the concentration of phosphorylated CheY protein is reduced. As a consequence, less phosphorylated CheY will bind to the motor complex and the tumbling frequency is reduced. The net result is a biased random walk in which the bacterium tends to move toward the direction in which the gradient of attractant concentration increases.

A simple model for the molecular control system that regulates chemotaxis is shown in Figure 5.15a. We start with the basic sensing and actuation mechanisms. A membrane-bound protein MCP (methyl-accepting chemotaxis protein) that is capable of binding to the external ligand serves as a signal transducing element from the cell exterior to the cytoplasm. Two other proteins, CheW and CheA, form a complex with MCP. This complex can either be in an active or inactive state. In the active state, CheA is autophosphorylated and serves as a phosphotransferase for two additional proteins, CheB and CheY. The phosphorylated form of CheY then binds to the motor complex, causing clockwise rotation of the motor (tumble).

The activity of the receptor complex is governed by two primary factors: the binding of a ligand molecule to the MCP protein and the presence or absence of up to four methyl groups on the MCP protein. The specific dependence on each of these factors is somewhat complicated. Roughly speaking, when the ligand L is bound to the receptor then the complex is less likely to be active. Furthermore, as more methyl groups are present, the ligand binding probability increases, allowing the gain of the sensor to be adjusted through methylation. Finally, even in the absence of ligand the receptor complex can be active, with the probability of it being active increasing with increased methylation. Figure 5.15b summarizes the possible states, their free energies and the probability of activity.

Several other elements are contained in the chemotaxis control circuit. The most important of these are implemented by the proteins CheR and CheB, both of which affect the receptor complex. CheR, which is constitutively produced in the cell, methylates the receptor complex at one of the four different methylation sites. Conversely, the phosphorylated form of CheB demethylates the receptor complex. As described above, the methylation patterns of the receptor complex affect its activity, which affects the phosphorylation of CheA and, in turn, phosphorylation of CheY and CheB. The combination of CheA, CheB and the methylation of the receptor complex forms a negative feedback loop: if the receptor is active, then CheA phosphorylates CheB, which in turn demethylates the receptor complex, making it less active. As we shall see when we investigate the detailed dynamics below, this feedback loop corresponds to a type of integral feedback law. This integral action allows the cell to adjust to different levels of ligand concentration, so that the behavior of the system is invariant to the absolute ligand levels.

Modeling

From a high level, we can view the chemotaxis as a dynamical system that takes the ligand L concentration as an input and produces the phosphorylated CheY bound to the motor complex as an output, which determines the tumbling frequency. We let T represent the receptor complex and we write A, B, Y and Z for CheA, CheB, CheY and CheZ, respectively. As in previous chapters, for a protein X we let X^* represent its phosphorylated form.

Each receptor complex can have multiple methyl groups attached and the activity of the receptor complex depends on both the amount of methylation and whether a ligand is attached to the receptor site (Figure 5.15b). Furthermore, the binding probabilities for the receptor also depend on the methylation pattern. We let T_i^x represent a receptor that has i methylation sites filled and ligand state x (which can be either u if unoccupied or o if occupied). We let m represent the maximum number of methylation sites ($m = 4$ for *E. coli*). Using this notation, the transitions between the states correspond to the following reactions (also shown in Figure 5.16):

$$T_i^u + L \rightleftharpoons T_i^o,$$

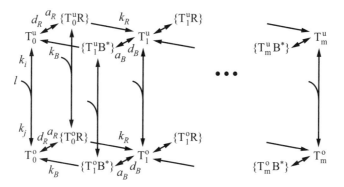

Figure 5.16: Methylation model for chemotaxis. Figure adapted from [9] (Box 1).

$$T_i^x + B^* \rightleftharpoons T_i^x{:}B^* \longrightarrow T_{i-1}^x + B^*, \qquad i > 0,\, x \in \{u, o\},$$
$$T_i^x + R \rightleftharpoons T_i^x{:}R \longrightarrow T_{i+1}^x + R, \qquad i < m,\, x \in \{u, o\},$$

in which the first reaction models the binding of the ligand L to the receptor complex, the second reaction models the demethylation of methylated receptor complex by B^*, and the third reaction models the methylation of the receptor complex by R.

We now must write reactions for each of the receptor complexes with CheA. Each form of the receptor complex has a different activity level that determines the extent to which CheA can be phosphorylated. Therefore, we write a separate reaction for each form, which for simplicity we assume to be a one-step process for each T_i^o and T_i^u species:

$$T_i^x + A \xrightarrow{k_i^x} T_i^x + A^*,$$

where $x \in \{o, u\}$ and $i = 0, \ldots, m$. As a consequence, the production rate of A^* by all the above reactions is given by

$$A \sum_{i=1}^{4} \left(k_i^o T_i^o + k_i^u T_i^u \right).$$

Considering that the ligand-receptor binding reaction is at its quasi-steady state because it is very fast, and letting $T_i = T_i^u + T_i^o$ represent the total amount of receptors with i sites methylated, we further have that

$$T_i^u = \frac{1}{1 + L/K_L} T_i, \qquad T_i^o = \frac{L/K_L}{1 + L/K_L} T_i,$$

in which K_L is the dissociation constant of the receptor complex-ligand binding. It follows that if we let T_i^A denote the "effective" concentration of T_i that phosphory-

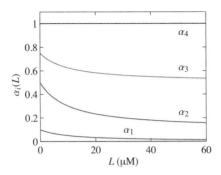

Figure 5.17: Probability of activity $\alpha_i(L)$ as a function of the ligand concentration L.

lates A, we have

$$k_i^a T_i^A = k_i^o T_i^o + k_i^u T_i^u = T_i \left(\frac{k_i^u + k_i^o (L/K_L)}{1 + L/K_L} \right),$$

for a suitably defined constant k_i^a. Hence, we can write

$$T_i^A = \alpha_i(L) T_i, \quad \text{with} \quad \alpha_i(L) = \frac{\alpha_i^o (L/K_L)}{1 + L/K_L} + \frac{\alpha_i^1}{1 + L/K_L},$$

and $T_i^I = T_i - T_i^A$, for suitable constants α_i^1 and α_i^o. The coefficients α_i^o and α_i^1 capture the effect of presence or absence of the ligand on the activity level of the complex and $\alpha_i(L)$ can be interpreted as the probability of activity. Following [83], we take the coefficients to be

$$\alpha_0^1 = 0, \qquad \alpha_1^1 = 0.1, \qquad \alpha_2^1 = 0.5, \qquad \alpha_3^1 = 0.75, \qquad \alpha_4^1 = 1,$$
$$\alpha_0^o = 0, \qquad \alpha_1^o = 0, \qquad \alpha_2^o = 0.1, \qquad \alpha_3^o = 0.5, \qquad \alpha_4^o = 1,$$

and choose $K_L = 10\ \mu\text{M}$. Figure 5.17 shows how each $\alpha_i(L)$ varies with L.

The total concentration of active receptors can now be written in terms of the receptor complex concentrations T_i and the activity probabilities $\alpha_i(L)$. We write the concentration of activated complex T^A and inactivated complex T^I as

$$T^A = \sum_{i=0}^{4} \alpha_i(L) T_i, \qquad T^I = \sum_{i=0}^{4} (1 - \alpha_i(L)) T_i.$$

These formulas can now be used in our dynamics as an effective concentration of active or inactive receptors. In particular, letting k_0 represent a lumped reaction rate constant for the phosphorylation of A by T^A, for which we employ a one-step reaction model, we have

$$\text{T}^A + \text{A} \xrightarrow{k_0} \text{T}^A + \text{A}^*. \tag{5.14}$$

We next model the transition between the methylation patterns on the receptor. For this, we assume that CheR binds only inactive receptors and phosphorylated CheB binds only to active receptors [83, 72]. This leads to the following reactions:

$$T_i^A + B^* \underset{d_B}{\overset{a_B}{\rightleftharpoons}} T_i^A{:}B^* \overset{k_B}{\longrightarrow} T_{i-1}^A + B^*, \qquad i \in \{2,3,4\},$$

$$T_1^A + B^* \underset{d_B}{\overset{a_B}{\rightleftharpoons}} T_1^A{:}B^*,$$

$$T_i^I + R \underset{d_R}{\overset{a_R}{\rightleftharpoons}} T_i^I{:}R \overset{k_R}{\longrightarrow} T_{i+1}^I + R, \qquad i \in \{1,2,3\},$$

$$T_4^I + R \underset{d_R}{\overset{a_R}{\rightleftharpoons}} T_4^I{:}R,$$

in which we accounted for the fact that R can still bind to T_4^I even without methylating it and B^* can still bind T_1^A even without demethylating it. Assuming the complexes' concentrations are at their quasi-steady state values, and letting $K_R = (d_R + k_R)/a_R$ and $K_B = (d_B + k_B)/a_B$ denote the Michaelis-Menten constant for the above enzymatic reactions, we have the following relations:

$$[T_i^R{:}R] = \frac{T_i^I R}{K_R}, \qquad [T_i^A{:}B^*] = \frac{T_i^A B^*}{K_B},$$

in which we have approximated d_R/a_R (d_B/a_B) by K_R (K_B) accounting for the fact that $k_R \ll d_R$ $(k_B \ll d_B)$. We can now write the differential equation for T_i considering the fact that

$$\frac{dT_i}{dt} = \frac{dT_i^I}{dt} + \frac{dT_i^A}{dt}.$$

Specifically, we have that

$$\frac{dT_i^I}{dt} = -k_R \frac{T_i^I R}{K_R} + k_R \frac{T_{i-1}^I R}{K_R},$$

$$\frac{dT_i^A}{dt} = -k_B \frac{T_i^A B^*}{K_B} + k_B \frac{T_{i+1}^A B^*}{K_B},$$

with the conservation laws $R_{\text{tot}} = R + \sum_{i=1}^4 [T_i^I{:}R]$ and $B_{\text{tot}}^* = B^* + \sum_{i=1}^4 [T_i{:}B^*]$, in which B_{tot}^* represents the total amount of phosphorylated CheB protein. Considering the quasi-steady state expressions for the complexes and the fact that $\sum_{i=1}^4 T_i^A = T^A$ and $\sum_{i=1}^4 T_i^I = T^I$, these conservation laws lead to

$$R = \frac{R_{\text{tot}}}{1 + T^I/K_R}, \qquad B^* = \frac{B_{\text{tot}}^*}{1 + T^A/K_B}.$$

Defining

$$r_R = k_R \frac{R_{\text{tot}}}{K_R + T^I}, \qquad r_B = k_B \frac{B_{\text{tot}}^*}{K_B + T^A},$$

which represent the effective rates of the methylation and demethylation reactions, we finally obtain that

$$\frac{d}{dt}T_i = r_R(1 - \alpha_{i-1}(L))T_{i-1} + r_B\alpha_{i+1}(L)T_{i+1} - r_R(1 - \alpha_i(L))T_i - r_B\alpha_i(L)T_i,$$

where the first and second terms represent transitions into this state via methylation or demethylation of neighboring states (see Figure 5.16) and the last two terms represent transitions out of the current state by methylation and demethylation, respectively. Note that the equations for T_0 and T_4 are slightly different since the demethylation and methylation reactions are not present, respectively.

Finally, we write the phosphotransfer and dephosphorylation reactions among CheA, CheB, and CheY, and the binding of CheY* to the motor complex M:

$$A^* + Y \xrightarrow{k_1} A + Y^*, \quad Y^* \xrightarrow{k_2} Y,$$

$$A^* + B \xrightarrow{k_3} A + B^*, \quad B^* \xrightarrow{k_4} B,$$

$$T_i^A{:}B^* \xrightarrow{k_4} T_i^A{:}B, \ i \in \{1,2,3,4\},$$

$$Y^* + M \underset{d}{\overset{a}{\rightleftharpoons}} Y^*{:}M,$$

in which the first reaction is the phosphotransfer from CheA* to CheY*, the second reaction is the dephosphorylation of CheY* by CheZ, the third reaction is the phosphotransfer from CheA* to CheB, the fourth and fifth reactions are the dephosphorylation of CheB*, which can be dephosphorylated also when bound to the receptor, and the last reaction is the binding of CheY* with the motor complex M to form the complex $Y^* : M$, which increases the probability that the motor will rotate clockwise. The resulting ODE model is given by

$$\frac{d}{dt}A^* = k_0 T^A A - k_1 A^* Y - k_3 A^* B,$$

$$\frac{d}{dt}Y^* = k_1 A^* Y - k_2 Y^* - aMY^* + d[\text{M}{:}Y^*],$$

$$\frac{d}{dt}B^*_{\text{tot}} = k_3 A^* B - k_4 B^*_{\text{tot}},$$

$$\frac{d}{dt}[\text{M}{:}Y^*] = aMY^* - d[\text{M}{:}Y^*],$$

with conservation laws

$$A + A^* = A_{\text{tot}}, \qquad B + B^*_{\text{tot}} = B_{\text{tot}},$$

$$Y + Y^* + [\text{M}{:}Y^*] = Y_{\text{tot}}, \qquad M + [\text{M}{:}Y^*] = M_{\text{tot}}.$$

Figure 5.18a shows the concentration of the phosphorylated proteins based on a simulation of the model. Initially, all species are started in their unphosphory-

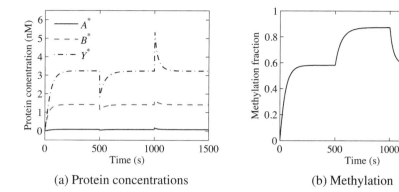

(a) Protein concentrations   (b) Methylation

Figure 5.18: Simulation and analysis of reduced-order chemotaxis model. The parameters are taken from Rao et al. [83] and given by $k_0 = 50\,\mathrm{s}^{-1}\mathrm{nM}^{-1}$, $k_1 = 100\,\mathrm{s}^{-1}\mathrm{nM}^{-1}$, $k_3 = 30\,\mathrm{s}^{-1}\mathrm{nM}^{-1}$, $k_2 = 30\,\mathrm{s}^{-1}\mathrm{nM}^{-1}$, $a = 50\,\mathrm{s}^{-1}\mathrm{nM}^{-1}$, $d = 19\,\mathrm{s}^{-1}$, and $k_4 = 1\,\mathrm{s}^{-1}$. The concentrations are $R_{\mathrm{tot}} = 0.2$ nM, $\sum_{i=0}^{4} T_i = 5$ nM, $A_{\mathrm{tot}} = 5$ nM, $B_{\mathrm{tot}} = 2$ nM, and $Y_{\mathrm{tot}} = 17.9$ nM. Also, we have $k_B = 0.5\,\mathrm{s}^{-1}$, $K_B = 5.5$ nM, $k_R = 0.255\,\mathrm{s}^{-1}$, and $K_R = 0.251$ nM.

lated and demethylated states. At time 500 s the ligand concentration is increased to $L = 10\,\mu$M and at time 1000 s it is returned to zero. We see that immediately after the ligand is added, the CheY* concentration drops, allowing longer runs between tumble motions. After a short period, however, the CheY* concentration adapts to the higher concentration and the nominal run versus tumble behavior is restored. Similarly, after the ligand concentration is decreased the concentration of CheY* increases, causing a larger fraction of tumbles (and subsequent changes in direction). Again, adaptation over a longer time scale returns the CheY concentration to its nominal value. The chemotaxis circuit pathway from T to CheY (to the motor) explains the sudden drop of CheY* when the ligand L is added. The slower methylation of the receptor complex catalyzed by CheR and removed by CheB slowly takes the value of CheY* to the pre-stimulus level.

Figure 5.18b helps explain the adaptation response. We see that the average amount of methylation of the receptor proteins increases when the ligand concentration is high, which decreases the activity of CheA (and hence decreases the phosphorylation of CheY).

Integral action

The perfect adaptation mechanism in the chemotaxis control circuitry has the same function as the use of integral action in control system design: by including a feedback on the integral of the error, it is possible to provide exact cancellation to constant disturbances. In this section we demonstrate that a simplified version of the dynamics can indeed be regarded as integral action of an appropriate signal. This interpretation was first pointed out by Yi et al. [102].

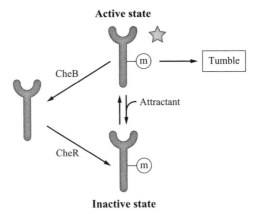

Figure 5.19: Reduced-order model of receptor activity. Star indicates activated complex and "m" indicates methylated complex. Figure adapted from [4], Figure 7.9.

We begin by formulating an even simpler model for the system dynamics that captures the basic features required to understand the integral action. We consider the receptor complex T and the kinase CheA as a single entity denoted by X, which can be either methylated or not. We ignore the number of methylated sites and simply group all the methylated forms into a lumped stated called X_m. Also, we assume that only the methylated state can be activated and the activity is determined by the ligand L concentration (through the functions $\alpha_i(L)$). We let X_m^* represent this active state and ignore the additional phosphorylation dynamics of CheY, so that we take the concentration X_m^* as our measure of overall activity.

We take the ligand into account by assuming that the transition between the active form X_m^* and the inactive form X_m depends on the ligand concentration: higher ligand concentration will increase the rate of transition to the inactive state. The activation/deactivation reactions are then written as

$$\text{R0:} \quad X_m^* \underset{k^r}{\overset{k^f(L)}{\rightleftharpoons}} X_m \quad \text{activation/deactivation,}$$

in which the deactivation rate $k^f(L)$ is an increasing function of L. As before, CheR methylates the receptor and CheB* demethylates it. We simplify the picture by only allowing CheB* to act on the active state X_m^* and CheR to act on the inactive state. Figure 5.19 shows the transitions between the various forms X.

This model is a considerable simplification from the ligand binding model that is illustrated in Figures 5.15b and 5.16. In the previous models, there is some probability of activity with or without methylation and with or without ligand. In this simplified model, we assume that only three states are of interest: demethylated, methylated/inactive and methylated/active. We also modify the way that ligand binding is captured and instead of keeping track of all of the possibilities in Figure 5.15b, we assume that the ligand transitions us from an active state X_m^* to an inactive X_m. These states and transitions are roughly consistent with the different energy levels and probabilities in Figure 5.15b, but it is clearly a much coarser model.

Accepting these approximations, the model illustrated in Figure 5.19 results in the methylation and demethylation reactions

$$R1: \quad X + R \rightleftharpoons X{:}R \longrightarrow X_m + R \qquad \text{methylation,}$$

$$R2: \quad X_m^* + B^* \rightleftharpoons X_m^*{:}B^* \longrightarrow X + B^* \quad \text{demethylation.}$$

For simplicity we take both R and B^* to have constant concentration.

We can further approximate the first and second reactions by their Michaelis-Menten forms, which yield net methylation and demethylation rates (for those reactions)

$$v_+ = k_R R \frac{X}{K_X + X}, \qquad v_- = k_B B^* \frac{X_m^*}{K_{X_m^*} + X_m^*}.$$

If we further assume that $X \gg K_X > 1$, then the methylation rate can be further simplified:

$$v_+ = k_R R \frac{X}{K_X + X} \approx K_R R.$$

Using these approximations, we can write the resulting dynamics for the overall system as

$$\frac{d}{dt} X_m = k_R R + k^f(L) X_m^* - k^r X_m,$$

$$\frac{d}{dt} X_m^* = -k_B B^* \frac{X_m^*}{K_{X_m^*} + X_m^*} - k^f(L) X_m^* + k^r X_m.$$

We wish to use this model to understand how the steady state activity level X_m^* depends on the ligand concentration L (which enters through the deactivation rate $k^f(L)$).

It will be useful to rewrite the dynamics in terms of the activated complex concentration X_m^* and the *total* methylated complex concentration $X_{m,\text{tot}} = X_m + X_m^*$. A simple set of algebraic manipulations yields

$$\frac{dX_m^*}{dt} = k^r(X_{m,\text{tot}} - X_m^*) - k_B B^* \frac{X_m^*}{K_{X_m^*} + X_m^*} - k^f(L) X_m^*,$$

$$\frac{dX_{m,\text{tot}}}{dt} = k_R R - k_B B^* \frac{X_m^*}{K_{X_m^*} + X_m^*}.$$

From the second equation, we see that the the concentration of methylated complex $X_{m,\text{tot}}$ is a balance between the action of the methylation reaction (R1, at rate v_+) and the demethylation reaction (R2, at rate v_-). Since the action of a ligand binding to the receptor complex increases the rate of deactivation of the complex (R0), in the presence of a ligand we will increase the amount of methylated complex and, (via reaction R1) eventually restore the amount of the activated complex. This represents the adaptation mechanism in this simplified model.

To further explore the effect of adaptation, we compute the equilibrium points for the system. Setting the time derivatives to zero, we obtain

$$X_{m,e}^* = \frac{K_{X_m^*} k_R R}{k_B B^* - k_R R},$$

$$X_{m,tot,e} = \frac{1}{k^r} \left(k^r X_m^* + k_B B^* \frac{X_m^*}{K_{X_m^*} + X_m^*} + k^f(L) X_m^* \right).$$

Note that the solution for the active complex $X_{m,e}^*$ in the first equation does not depend on $k^f(L)$ (or k^r) and hence the steady state solution is independent of the ligand concentration. Thus, in steady state, the concentration of activated complex adapts to the steady state value of the ligand that is present, making it insensitive to the steady state value of this input.

In order to demonstrate that after a perturbation due to addition of the ligand the value of X_m^* returns to its equilibrium, we need to prove that the equilibrium point $(X_{m,tot,e}, X_{m,e}^*)$ is asymptotically stable. To do so, let $x = X_{m,tot}$, $y = X_m^*$ and rewrite the system model as

$$\frac{dx}{dt} = k_R R - k_B B^* \frac{y}{K_{X_m^*} + y},$$

$$\frac{dy}{dt} = k^r(x - y) - k_B B^* \frac{y}{K_{X_m^*} + y} - k^f(L) y,$$

which is in the standard integral feedback form introduced in Section 3.2. The stability of the equilibrium point $(x_e, y_e) = (X_{m,tot,e}, X_{m,e}^*)$ can be determined by calculating the Jacobian matrix J of this system at the equilibrium. This gives

$$J = \begin{pmatrix} 0 & -k_B B^* \frac{K_{X_m^*}}{(y_e + K_{X_m^*})^2} \\ k^r & -k_B B^* \frac{K_{X_m^*}}{(y_e + K_{X_m^*})^2} - k^f(L) - k^r \end{pmatrix},$$

for which

$$\operatorname{tr}(J) < 0 \qquad \text{and} \qquad \det(J) > 0,$$

implying that the equilibrium point is asymptotically stable.

The dynamics for $X_{m,tot}$ can be viewed as an integral action: when the concentration of X_m^* matches its reference value (with no ligand present), the quantity of methylated complex $X_{m,tot}$ remains constant. But if $X_{m,tot}$ does not match this reference value, then $X_{m,tot}$ increases at a rate proportional to the methylation "error" (measured here by difference in the nominal reaction rates v_+ and v_-). It can be shown that this type of integral action is necessary to achieve perfect adaptation in a robust manner [102].

Exercises

5.1 Consider the negatively autoregulated system given in equation (5.4). Determine the frequency response with respect to noise input Γ_2 and compare its magnitude to that of the open loop system in equation (5.2).

5.2 Consider the contribution of the negative autoregulation given by the parameter g in equation (5.6). Study how g changes when the value of the dissociation constant K_d is changed.

5.3 Consider the negatively autoregulated system

$$\frac{dA}{dt} = \frac{\beta}{1 + (A/K)^n} - \gamma A.$$

Explore through linearization how increasing the Hill coefficient affects the response time of the system. Also, compare the results of the linearization analysis to the behavior of the nonlinear system obtained through simulation.

5.4 Consider the toggle switch model

$$\frac{dA}{dt} = \frac{\beta_A}{1 + (B/K)^n} - \gamma A, \qquad \frac{dB}{dt} = \frac{\beta_B}{1 + (A/K)^m} - \gamma B.$$

Here, we are going to explore the parameter space that makes the system work as a toggle switch. To do so, answer the following questions:

(i) Consider $m = n = 1$. Determine the number and stability of the equilibria as the values of β_A and β_B are changed.

(ii) Consider $m = 1$ and $n > 1$ and determine the number and stability of the equilibria (as other parameters change).

(iii) Consider $m = n = 2$. Determine parameter conditions on $\beta_A, \beta_B, \gamma, K$ for which the system is bistable, i.e., there are two stable steady states.

5.5 Consider the repressilator model and the parameter space for oscillations provided in Figure 5.5. Determine how this parameter space changes if the value of K in the Hill function is changed.

5.6 Consider the "generalized" model of the repressilator in which we have m repressors (with m an odd number) in the loop. Explore via simulation the fact that when m is increased, the system oscillates for smaller values of the Hill coefficient n.

5.7 Consider the activator-repressor clock model given in equations (5.11). Determine the number and stability of the equilibria as performed in the text for the case in which $m > 1$.

5.8 Consider the feedforward circuit shown in Figure 5.11. Assume that we account for cooperativity such that the model becomes

$$\frac{dA}{dt} = k_0 u - \gamma A, \qquad \frac{dB}{dt} = \frac{k_1 u}{1 + (A/K_d)^n} - \gamma B.$$

Show that the adaptation property still holds under suitable parameter conditions.

Chapter 6
Interconnecting Components

In Chapter 2 and Chapter 5 we studied the behavior of simple biomolecular modules, such as oscillators, toggles, self-repressing circuits, signal transduction and amplification systems, based on reduced-order models. One natural step forward is to create larger and more complex systems by composing these modules together. In this chapter, we illustrate problems that need to be overcome when interconnecting components and propose a number of engineering solutions based on the feedback principles introduced in Chapter 3. Specifically, we explain how loading effects arise at the interconnection between modules, which change the expected circuit behavior. These loading problems appear in several other engineering domains, including electrical, mechanical, and hydraulic systems, and have been largely addressed by the respective engineering communities. In this chapter, we explain how similar engineering solutions can be employed in biomolecular systems to defeat loading effects and guarantee "modular" interconnection of circuits. In Chapter 7, we further study loading of the cellular environment by synthetic circuits employing the same framework developed in this chapter.

6.1 Input/output modeling and the modularity assumption

The input/output modeling introduced in Chapter 1 and further developed in Chapter 3 has been employed so far to describe the behavior of various modules and subsystems. This input/output description of a system allows us to connect systems together by setting the input u_2 of a downstream system equal to the output y_1 of the upstream system (Figure 6.1) and has been extensively used in the previous

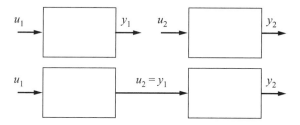

Figure 6.1: In the input/output modeling framework, systems are interconnected by statically assigning the value of the output of the upstream system to the input of the downstream system.

chapters.

Each node of a gene circuit (see Figure 5.1, for example), has been modeled as an input/output system taking the concentrations of transcription factors as input and giving, through the processes of transcription and translation, the concentration of another transcription factor as an output. For example, node C in the repressilator has been modeled as a second-order system that takes the concentration of transcription factor B as an input through the Hill function and gives transcription factor C as an output. This is of course not the only possible choice for decomposing the system. We could in fact let the mRNA or the RNA polymerase flowing along the DNA, called PoPS (polymerase per second) [28], play the role of input and output signals. Similarly, a signal transduction network is usually composed of protein covalent modification modules, which take a modifying enzyme (a kinase in the case of phosphorylation) as an input and gives the modified protein as an output.

This input/output modeling framework is extremely useful because it allows us to predict the behavior of an interconnected system from the behavior of the isolated modules. For example, the location and number of equilibria in the toggle switch of Section 5.3 were predicted by intersecting the steady state input/output characteristics, determined by the Hill functions, of the isolated modules A and B. Similarly, the number of equilibria in the repressilator of Section 5.4 was predicted by modularly composing the input/output steady state characteristics, again determined by the Hill functions, of the three modules composing the circuit. Finally, criteria for the existence of a limit cycle in the activator-repressor clock of Section 5.5 were based on comparing the speed of the activator module's dynamics to that of the repressor module's dynamics.

For this input/output interconnection framework to reliably predict the behavior of connected modules, it is necessary that the input/output (dynamic) behavior of a system does not change upon interconnection to another system. We refer to the property by which a system input/output behavior does not change upon interconnection as *modularity*. All the designs and models described in the previous chapter assume that the modularity property holds. In this chapter, we question this assumption and investigate when modularity holds in gene and in signal transduction circuits. Further, we illustrate design methods, based on the techniques of Chapter 3, to create functionally modular systems.

6.2   Introduction to retroactivity

The modularity assumption implies that when two modules are connected together, their behavior does not change because of the interconnection. However, a fundamental systems engineering issue that arises when interconnecting subsystems is how the process of transmitting a signal to a "downstream" component affects the dynamic state of the upstream sending component. This issue, the effect of "loads"

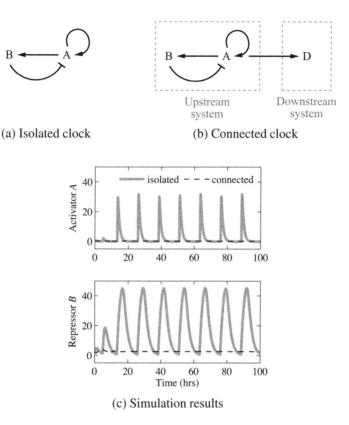

(a) Isolated clock   (b) Connected clock

(c) Simulation results

Figure 6.2: Interconnection of an activator-repressor clock to a downstream system. (a) The activator-repressor clock is isolated. (b) The clock is connected to a downstream system. (c) When the clock is connected, periodic behavior of the protein's concentration is lost and oscillations are quenched. The clock hence fails to transmit the desired periodic stimulation to the downstream system. In all simulations, we have chosen the parameters of the clock as in Figure 5.9. For the system in (b), we added the reversible binding reaction of A with sites p in the downstream system: $nA + p \rightleftharpoons C$ with conservation law $p_{tot} = p + C$, with $p_{tot} = 5nM$, association rate constant $k_{on} = 50 \text{ min}^{-1} \text{ nM}^{-n}$, and dissociation rate constant $k_{off} = 50 \text{ min}^{-1}$ (see Exercise 6.12).

on the output of a system, is well-understood in many engineering fields such as electrical engineering. It has often been pointed out that similar issues may arise for biological systems. These questions are especially delicate in design problems, such as those described in Chapter 5.

For example, consider a biomolecular clock, such as the activator-repressor clock introduced in Section 5.5 and shown in Figure 6.2a. Assume that the activator protein concentration $A(t)$ is now used as a communicating species to synchronize or provide the timing to a downstream system D (Figure 6.2b). From a systems/signals point of view, $A(t)$ becomes an *input* to the downstream system D. The terms "upstream" and "downstream" reflect the direction in which we think of

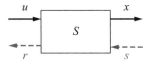

Figure 6.3: A system S input and output signals. The r and s signals denote signals origi-nating by retroactivity upon interconnection [22].

signals as traveling, *from* the clock *to* the systems being synchronized. However, this is only an idealization because when A is taken as an input by the downstream system it binds to (and unbinds from) the promoter that controls the expression of D. These additional binding/unbinding reactions compete with the biochemical interactions that constitute the upstream clock and may therefore disrupt the opera-tion of the clock itself (Figure 6.2c). We call this back-effect *retroactivity* to extend the notion of impedance or loading to non-electrical systems and in particular to biomolecular systems. This phenomenon, which in principle may be used in an advantageous way by natural systems, can be deleterious when designing synthetic systems.

One possible approach to avoid disrupting the behavior of the clock is to in-troduce a gene coding for a new protein X, placed under the control of the same promoter as the gene for A, and using the concentration of X, which presumably mirrors that of A, to drive the downstream system. However, this approach still has the problem that the concentration of X may be altered and even disrupted by the addition of downstream systems that drain X, as we shall see in the next section. The net result is that the downstream systems are not properly timed as X does not transmit the desired signal.

To model a system with retroactivity, we add to the input/output modeling framework used so far an additional input, called s, to account for any change that may occur upon interconnection with a downstream system (Figure 6.3). That is, s models the fact that whenever y is taken as an input to a downstream sys-tem the value of y may change, because of the physics of the interconnection. This phenomenon is also called in the physics literature "the observer effect," implying that no physical quantity can be measured without being altered by the measure-ment device. Similarly, we add a signal r as an additional output to model the fact that when a system is connected downstream of another one, it will send a signal upstream that will alter the dynamics of that system. More generally, we define a system S to have internal state x, two types of inputs, and two types of outputs: an input "u," an output "y" (as before), a *retroactivity to the input* "r," and a *retroac-tivity to the output* "s." We will thus represent a system S by the equations

$$\frac{dx}{dt} = f(x,u,s), \qquad y = h(x,u,s), \qquad r = R(x,u,s), \qquad (6.1)$$

where f, g, and R are arbitrary functions and the signals x, u, s, r, and y may be

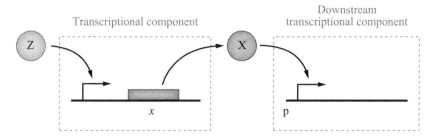

Figure 6.4: A transcriptional component takes protein concentration Z as input u and gives protein concentration X as output y. The downstream transcriptional component takes protein concentration X as its input.

scalars or vectors. In such a formalism, we define the input/output model of the *isolated system* as the one in equation (6.1) without r in which we have also set $s = 0$.

Let S_i be a system with inputs u_i and s_i and with outputs y_i and r_i. Let S_1 and S_2 be two systems with disjoint sets of internal states. We define the interconnection of an upstream system S_1 with a downstream system S_2 by simply setting $y_1 = u_2$ and $s_1 = r_2$. For interconnecting two systems, we require that the two systems do not have internal states in common.

It is important to note that while retroactivity s is a back action from the downstream system to the upstream one, it is conceptually different from feedback. In fact, retroactivity s is nonzero any time y is transmitted to the downstream system. That is, it is not possible to send signal y to the downstream system without retroactivity s. By contrast, feedback from the downstream system can be removed even when the upstream system sends signal y.

6.3   Retroactivity in gene circuits

In the previous section, we have introduced retroactivity as a general concept modeling the fact that when an upstream system is input/output connected to a downstream one, its behavior can change. In this section, we focus on gene circuits and show what form retroactivity takes and what its effects are.

Consider the interconnection of two transcriptional components illustrated in Figure 6.4. A transcriptional component is an input/output system that takes the transcription factor concentration Z as input and gives the transcription factor concentration X as output. The activity of the promoter controlling gene x depends on the amount of Z bound to the promoter. If $Z = Z(t)$, such an activity changes with time and, to simplify notation, we denote it by $k(t)$. We assume here that the mRNA dynamics are at their quasi-steady state. The reader can verify that all the results hold unchanged when the mRNA dynamics are included (see Exercise 6.1).

We write the dynamics of X as

$$\frac{dX}{dt} = k(t) - \gamma X, \tag{6.2}$$

in which γ is the decay rate constant of the protein. We refer to equation (6.2) as the *isolated system* dynamics.

Now, assume that X drives a downstream transcriptional module by binding to a promoter p (Figure 6.4). The reversible binding reaction of X with p is given by

$$X + p \underset{k_{off}}{\overset{k_{on}}{\rightleftharpoons}} C,$$

in which C is the complex protein-promoter and k_{on} and k_{off} are the association and dissociation rate constants of protein X to promoter site p. Since the promoter is not subject to decay, its total concentration p_{tot} is conserved so that we can write $p + C = p_{tot}$. Therefore, the new dynamics of X are governed by the equations

$$\frac{dX}{dt} = k(t) - \gamma X + [k_{off}C - k_{on}(p_{tot} - C)X], \qquad \frac{dC}{dt} = -k_{off}C + k_{on}(p_{tot} - C)X. \tag{6.3}$$

We refer to this system as the *connected system*. Comparing the rate of change of X in the connected system to that in the isolated system (6.2), we notice the additional rate of change $[k_{off}C - k_{on}(p_{tot} - C)X]$ of X in the connected system. Hence, we have

$$s = [k_{off}C - k_{on}(p_{tot} - C)X],$$

and $s = 0$ when the system is isolated. We can interpret s as a mass flow between the upstream and the downstream system, similar to a current in electrical circuits.

How large is the effect of retroactivity s on the dynamics of X and what are the biological parameters that affect it? We focus on the retroactivity to the output s as we can analyze the effect of the retroactivity to the input r on the upstream system by simply analyzing the dynamics of Z in the presence of the promoter regulating the expression of gene x.

The effect of retroactivity s on the behavior of X can be very large (Figure 6.5). By looking at Figure 6.5, we notice that the effect of retroactivity is to "slow down" the dynamics of $X(t)$ as the response time to a step input increases and the response to a periodic signal appears attenuated and phase-shifted. We will come back to this more precisely in the next section.

These effects are undesirable in a number of situations in which we would like an upstream system to "drive" a downstream one, for example, when a biological oscillator has to time a number of downstream processes. If, due to the retroactivity, the output signal of the upstream process becomes too low and/or out of phase with the output signal of the isolated system (as in Figure 6.5), the desired coordination between the oscillator and the downstream processes will be lost. We next provide a procedure to quantify the effect of retroactivity on the dynamics of the upstream system.

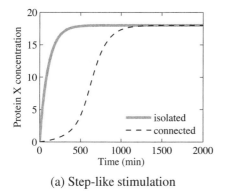
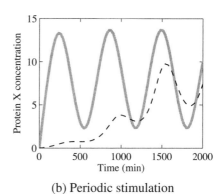

(a) Step-like stimulation   (b) Periodic stimulation

Figure 6.5: The effect of retroactivity. The solid line represents $X(t)$ originating by equation (6.2), while the dashed line represents $X(t)$ obtained by equations (6.3). Both transient and permanent behaviors are different. Here, $k(t) = 0.18$ in (a) and $k(t) = 0.01(8 + 8 sin(\omega t))$ with $\omega = 0.01$ min^{-1} in (b). The parameter values are given by $k_{on} = 10$ min^{-1}nM^{-1}, $k_{off} = 10$ min^{-1}, $\gamma = 0.01$ min^{-1}, and $p_{tot} = 100$ nM. The frequency of oscillations is chosen to have a period of about 11 hours in accordance to what is experimentally observed in the synthetic clock of [6].

Quantification of the retroactivity to the output

In this section, we provide a general approach to quantify the retroactivity to the output. To do so, we quantify the difference between the dynamics of X in the isolated system (6.2) and the dynamics of X in the connected system (6.3) by establishing conditions on the biological parameters that make the two dynamics close to each other. This is achieved by exploiting the difference of time scales between the protein production and decay processes and binding/unbinding reactions, mathematically described by $k_{off} \gg \gamma$. By virtue of this separation of time scales, we can approximate system (6.3) by a one-dimensional system describing the evolution of X on the slow manifold (see Section 3.5).

To this end, note that equations (6.3) are not in standard singular perturbation form: while C is a fast variable, X is neither fast nor slow since its differential equation includes both fast and slow terms. To explicitly model the difference of time scales, we let $z = X + C$ be the total amount of protein X (bound and free) and rewrite system (6.3) in the new variables (z, C). Letting $\epsilon = \gamma/k_{off}$, $K_d = k_{off}/k_{on}$, and $k_{on} = \gamma/(\epsilon K_d)$, system (6.3) can be rewritten as

$$\frac{dz}{dt} = k(t) - \gamma(z - C), \qquad \epsilon\frac{dC}{dt} = -\gamma C + \frac{\gamma}{K_d}(p_{tot} - C)(z - C), \qquad (6.4)$$

in which z is a slow variable. The reader can check that the slow manifold of system (6.4) is locally exponentially stable (see Exercise 6.2).

We can obtain an approximation of the dynamics of X in the limit in which ϵ is very small by setting $\epsilon = 0$. This leads to

$$-\gamma C + \frac{\gamma}{K_d}(p_{tot} - C)X = 0 \quad \Longrightarrow \quad C = g(X) \text{ with } g(X) = \frac{p_{tot}X}{X + K_d}.$$

Since $\dot{z} = \dot{X} + \dot{C}$, we have that $\dot{z} = \dot{X} + (dg/dX)\dot{X}$. This along with $\dot{z} = k(t) - \gamma X$ lead to

$$\frac{dX}{dt} = (k(t) - \gamma X)\left(\frac{1}{1 + dg/dX}\right). \tag{6.5}$$

The difference between the dynamics in equation (6.5) (the connected system after a fast transient) and the dynamics in equation (6.2) (the isolated system) is zero when the term $dg(X)/dX$ in equation (6.5) is zero. We thus consider the term $dg(X)/dX$ as a quantification of the retroactivity s after a fast transient in the approximation in which $\epsilon \approx 0$. We can also interpret the term $dg(X)/dX$ as a percentage variation of the dynamics of the connected system with respect to the dynamics of the isolated system at the quasi-steady state. We next determine the physical meaning of such a term by calculating a more useful expression that is a function of key biochemical parameters. Specifically, we have that

$$\frac{dg(X)}{dX} = \frac{p_{tot}/K_d}{(X/K_d + 1)^2} =: \mathcal{R}(X). \tag{6.6}$$

The retroactivity measure $\mathcal{R}$ is low whenever the ratio p_{tot}/K_d, which can be seen as an effective load, is low. This is the case if the affinity of the binding sites p is small (K_d large) or if p_{tot} is low. Also, the retroactivity measure is dependent on X in a nonlinear fashion and it is such that it is maximal when X is the smallest. The expression of $\mathcal{R}(X)$ provides an operative quantification of retroactivity: such an expression can be evaluated once the dissociation constant of X is known, the concentration of the binding sites p_{tot} is known, and X is also measured. From equations (6.5) and (6.6), it follows that the rate of change of X in the connected system is smaller than that in the isolated system, that is, retroactivity slows down the dynamics of the transcriptional system. This has also been experimentally reported in [51].

Summarizing, the modularity assumption introduced in Section 6.1 holds only when the value of $\mathcal{R}(X)$ is small enough. Thus, the design of a simple circuit can assume modularity if the interconnections among the composing modules can be designed so that the value of $\mathcal{R}(X)$ is low. When designing the system, this can be guaranteed by placing the promoter sites p on low copy number plasmids or even on the chromosome (with copy number equal to 1). High copy number plasmids are expected to lead to non-negligible retroactivity effects on X.

Note however that in the presence of very low affinity and/or very low amount of promoter sites, the amount of complex C will be very low. As a consequence, the amplitude of the transmitted signal to downstream systems may also be very small so that noise may become a bottleneck. A better approach may be to design insulation devices (as opposed to designing the interconnection for low retroactivity) to buffer systems from possibly large retroactivity as explained later in the chapter.

Effects of retroactivity on the frequency response

In order to explain the amplitude attenuation and phase lag due to retroactivity observed in Figure 6.5, we linearize the system about its equilibrium and determine the effect of retroactivity on the frequency response. To this end, consider the input in the form $k(t) = \bar{k} + A_0 \sin(\omega t)$. Let $X_e = \bar{k}/\gamma$ and $C_e = p_{\text{tot}} X_e/(X_e + K_d)$ be the equilibrium values corresponding to $\bar{k}$. The isolated system is already linear, so there is no need to perform linearization and the transfer function from k to X is given by

$$G_{Xk}^i(s) = \frac{1}{s + \gamma}.$$

For the connected system (6.5), let $(\bar{k}, X_e)$ denote the equilibrium, which is the same as for the isolated system, and let $\tilde{k} = k - \bar{k}$ and $x = X - X_e$ denote small perturbations about this equilibrium. Then, the linearization of system (6.5) about $(\bar{k}, X_e)$ is given by

$$\frac{dx}{dt} = (\tilde{k}(t) - \gamma x) \frac{1}{1 + (p_{\text{tot}}/K_d)/(X_e/K_d + 1)^2}.$$

Letting $\bar{R} := (p_{\text{tot}}/K_d)/(X_e/K_d + 1)^2$, we obtain the transfer function from $\tilde{k}$ to x of the connected system linearization as

$$G_{Xk}^c = \frac{1}{1 + \bar{R}} \frac{1}{s + \gamma/(1 + \bar{R})}.$$

Hence, we have the following result for the frequency response magnitude and phase:

$$M^i(\omega) = \frac{1}{\sqrt{\omega^2 + \gamma^2}}, \qquad \phi^i(\omega) = \tan^{-1}(-\omega/\gamma),$$

$$M^c(\omega) = \frac{1}{1 + \bar{R}} \frac{1}{\sqrt{\omega^2 + \gamma^2/(1 + \bar{R})^2}}, \qquad \phi^c(\omega) = \tan^{-1}(-\omega(1 + \bar{R})/\gamma),$$

from which one obtains that $M^i(0) = M^c(0)$ and, since $\bar{R} > 0$, the bandwidth of the connected system $\gamma/(1 + \bar{R})$ is lower than that of the isolated system γ. As a consequence, we have that $M^i(\omega) > M^c(\omega)$ for all $\omega > 0$. Also, the phase shift of the connected system is larger than that of the isolated system since $\phi^c(\omega) < \phi^i(\omega)$. This explains why the plots of Figure 6.5 show attenuation and phase shift in the response of the connected system.

When the frequency of the input stimulation $k(t)$ is sufficiently lower than the bandwidth of the connected system $\gamma/(1 + \bar{R})$, then the connected and isolated systems will respond similarly. Hence, the effects of retroactivity are tightly related to

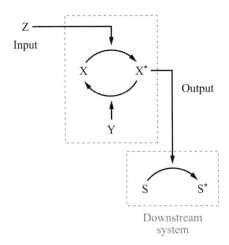

Figure 6.6: Covalent modification cycle with its input, output, and downstream system.

the time scale properties of the input signals and of the system, and mitigation of retroactivity is required only when the frequency range of the signals of interest is larger than the connected system bandwidth $\gamma/(1 + \bar{R})$ (see Exercise 6.4).

6.4   Retroactivity in signaling systems

Signaling systems are circuits that take external stimuli as inputs and, through a sequence of biomolecular reactions, transform them to signals that control how cells respond to their environment. These systems are usually composed of covalent modification cycles such as phosphorylation, methylation, and uridylylation, and connected in cascade fashion, in which each cycle has multiple downstream targets or substrates (refer to Figure 6.6). An example is the MAPK cascade, which we have analyzed in Section 2.5. Since covalent modification cycles always have downstream targets, such as DNA binding sites or other substrates, it is particularly important to understand whether and how retroactivity from these downstream systems affects the response of the upstream cycles to input stimulation. In this section, we study this question both for the steady state and transient response.

Steady state effects of retroactivity

We have seen in Section 2.4 that one important characteristic of signaling systems and, in particular, of covalent modification cycles, is the steady state input/output characteristic curve. We showed in Section 2.4 that when the Michaelis-Menten constants are sufficiently small compared to the amount of total protein, the steady state characteristic curve of the cycle becomes ultrasensitive, a condition called zero-order ultrasensitivity. When the cycle is connected to its downstream targets, this steady state characteristic curve changes. In order to understand how this happens, we rewrite the reaction rates and the corresponding differential equation model for the covalent modification cycle of Section 2.4, adding the binding of X* to its downstream target S. Referring to Figure 6.6, we have the following reactions:

$$Z + X \underset{d_1}{\overset{a_1}{\rightleftharpoons}} C_1 \overset{k_1}{\rightarrow} X^* + Z, \qquad Y + X^* \underset{d_2}{\overset{a_2}{\rightleftharpoons}} C_2 \overset{k_2}{\rightarrow} X + Y,$$

to which we add the binding reaction of X* with its substrate S:

$$X^* + S \underset{k_{\mathrm{off}}}{\overset{k_{\mathrm{on}}}{\rightleftharpoons}} C,$$

in which C is the complex of X* with S. In addition to this, we have the conservation laws $X_{\mathrm{tot}} = X^* + X + C_1 + C_2 + C$, $Z_{\mathrm{tot}} = Z + C_1$, and $Y_{\mathrm{tot}} = Y + C_2$.

The ordinary differential equations governing the system are given by

$$\frac{dC_1}{dt} = a_1 X Z - (d_1 + k_1) C_1,$$

$$\frac{dX^*}{dt} = -a_2 X^* Y + d_2 C_2 + k_1 C_1 - k_{\mathrm{on}} S X^* + k_{\mathrm{off}} C,$$

$$\frac{dC_2}{dt} = a_2 X^* Y - (d_2 + k_2) C_2,$$

$$\frac{dC}{dt} = k_{\mathrm{on}} X^* S - k_{\mathrm{off}} C.$$

The input/output steady state characteristic curve is found by solving this system for the equilibrium. In particular, by setting $\dot{C}_1 = 0$, $\dot{C}_2 = 0$, using that $Z = Z_{\mathrm{tot}} - C_1$ and that $Y = Y_{\mathrm{tot}} - C_2$, we obtain the familiar expressions for the complexes:

$$C_1 = \frac{Z_{\mathrm{tot}} X}{X + K_1}, \qquad C_2 = \frac{Y_{\mathrm{tot}} X^*}{X^* + K_2},$$

with

$$K_1 = \frac{d_1 + k_1}{a_1}, \qquad K_2 = \frac{d_2 + k_2}{a_2}.$$

By setting $\dot{X}^* + \dot{C}_2 + \dot{C} = 0$, we obtain $k_1 C_1 = k_2 C_2$, which leads to

$$V_1 \frac{X}{X + K_1} = V_2 \frac{X^*}{X^* + K_2}, \qquad V_1 = k_1 Z_{\mathrm{tot}} \quad \text{and} \quad V_2 = k_2 Y_{\mathrm{tot}}. \tag{6.7}$$

By assuming that the substrate X_{tot} is in excess compared to the enzymes, we have that $C_1, C_2 \ll X_{\mathrm{tot}}$ so that $X \approx X_{\mathrm{tot}} - X^* - C$, in which (from setting $\dot{C} = 0$) $C = X^* S / K_{\mathrm{d}}$ with $K_{\mathrm{d}} = k_{\mathrm{off}} / k_{\mathrm{on}}$, leading to $X \approx X_{\mathrm{tot}} - X^* (1 + S / K_{\mathrm{d}})$. Calling

$$\lambda = \frac{S}{K_{\mathrm{d}}},$$

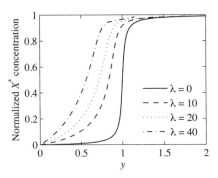

Figure 6.7: Effect of retroactivity on the steady state input/output characteristic curve of a covalent modification cycle. The addition of downstream target sites makes the input/output characteristic curve more linear-like, that is, retroactivity makes a switch-like response into a more graded response. The plot is obtained for $K_1/X_{\text{tot}} = K_2/X_{\text{tot}} = 0.01$ and the value of X^* is normalized to its maximum given by $X_{\text{tot}}/(1 + \lambda)$.

equation (6.7) finally leads to

$$y := \frac{V_1}{V_2} = \frac{X^*\left((K_1/(1+\lambda)) + ((X_{\text{tot}}/(1+\lambda)) - X^*)\right)}{(K_2 + X^*)((X_{\text{tot}}/(1+\lambda)) - X^*)}. \tag{6.8}$$

Here, we can interpret λ as an effective load, which increases with the amount of targets of X^* but also with the affinity of these targets $(1/K_d)$.

We are interested in how the shape of the steady state input/output characteristic curve of X^* changes as a function of y when the effective load λ is changed. As seen in Section 2.4, a way to quantify the sensitivity of the steady state characteristic curve is to calculate the response coefficient $R = y_{90}/y_{10}$. The maximal value of X^* obtained as $y \to \infty$ is given by $X_{\text{tot}}/(1 + \lambda)$. Hence, from equation (6.8), we have that

$$y_{90} = \frac{(\bar{K}_1 + 0.1)0.9}{(\bar{K}_2(1+\lambda) + 0.9)0.1}, \qquad y_{10} = \frac{(\bar{K}_1 + 0.9)0.1}{(\bar{K}_2(1+\lambda) + 0.1)0.9},$$

with

$$\bar{K}_1 := \frac{K_1}{X_{\text{tot}}}, \qquad \bar{K}_2 = \frac{K_2}{X_{\text{tot}}},$$

so that

$$R = 81\frac{(\bar{K}_1 + 0.1)(\bar{K}_2(1+\lambda) + 0.1)}{(\bar{K}_2(1+\lambda) + 0.9)(\bar{K}_1 + 0.9)}.$$

Comparing this expression with the one obtained in equation (2.31) for the isolated covalent modification cycle, we see that the net effect of the downstream target S is that of increasing the Michaelis-Menten constant K_2 by the factor $(1 + \lambda)$. Hence, we should expect that with increasing load, the steady state characteristic curve should be more linear-like. This is confirmed by the simulations shown in Figure 6.7 and it was also experimentally demonstrated in signal transduction circuits reconstituted in vitro [95].

One can check that R is a monotonically increasing function of λ. In particular, as λ increases, the value of R tends to $81(\bar{K}_1 + 0.1)/(\bar{K}_2 + 0.9)$, which, in turn, tends to 81 for $\bar{K}_1, \bar{K}_2 \to \infty$. When $\lambda = 0$, we recover the results of Section 2.4.

Dynamic effects of retroactivity

In order to understand the effects of retroactivity on the temporal response of a covalent modification cycle, we consider changes in Z_{tot} and analyze the temporal response of the cycle to these changes. To perform this analysis more easily, we seek a one-dimensional approximation of the X^* dynamics by exploiting time scale separation.

Specifically, we have that $d_i, k_{off} \gg k_1, k_2$, so we can choose $\epsilon = k_1/k_{off}$ as a small parameter and $w = X^* + C + C_2$ as a slow variable. By setting $\epsilon = 0$, we obtain $C_1 = Z_{tot}X/(X + K_1)$, $C_2 = Y_{tot}X^*/(X^* + K_2) =: g(X^*)$, and $C = \lambda X^*$, where Z_{tot} is now a time-varying input signal. Hence, the dynamics of the slow variable w on the slow manifold are given by

$$\frac{dw}{dt} = k_1 \frac{Z_{tot}(t)X}{X + K_1} - k_2 Y_{tot} \frac{X^*}{X^* + K_2}.$$

Using

$$\frac{dw}{dt} = \frac{dX^*}{dt} + \frac{dC}{dt} + \frac{dC_2}{dt}, \quad \frac{dC}{dt} = \lambda \frac{dX^*}{dt}, \quad \frac{dC_2}{dt} = \frac{\partial g}{\partial X^*} \frac{dX^*}{dt},$$

and the conservation law $X = X_{tot} - X^*(1 + \lambda)$, we finally obtain the approximated X^* dynamics as

$$\frac{dX^*}{dt} = \frac{1}{1 + \lambda} \left(k_1 \frac{Z_{tot}(t)(X_{tot} - X^*(1 + \lambda))}{(X_{tot} - X^*(1 + \lambda)) + K_1} - k_2 Y_{tot} \frac{X^*}{X^* + K_2} \right), \tag{6.9}$$

where we have assumed that $Y_{tot}/K_2 \ll S/K_d$, so that the effect of the binding dynamics of X^* with Y (modeled by $\partial g/\partial X^*$) is negligible with respect to λ. The reader can verify this derivation as an exercise (see Exercise 6.7).

From this expression, we can understand the effect of the load λ on the rise time and decay time in response to large step input stimuli Z_{tot}. For the decay time, we can assume an initial condition $X^*(0) \neq 0$ and $Z_{tot}(t) = 0$ for all t. In this case, we have that

$$\frac{dX^*}{dt} = -k_2 Y_{tot} \frac{X^*}{X^* + K_2} \frac{1}{1 + \lambda},$$

from which, since $\lambda > 0$, it follows that the transient will be slower than when $\lambda = 0$ and hence that the system will have an increased decay time due to retroactivity. For the rise time, one can assume $Z_{tot} \neq 0$ and $X^*(0) = 0$. In this case, at least initially we have that

$$(1 + \lambda) \frac{dX^*}{dt} = \left(k_1 \frac{Z_{tot}(X_{tot} - X^*(1 + \lambda))}{(X_{tot} - X^*(1 + \lambda)) + K_1} \right),$$

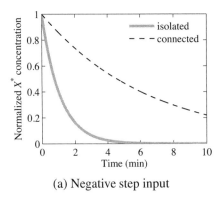

(a) Negative step input

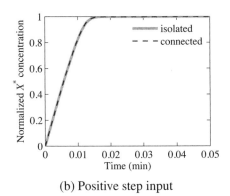

(b) Positive step input

Figure 6.8: Effect of retroactivity on the temporal response of a covalent modification cycle. (a) Response to a negative step. The presence of the load makes the response slower. (b) Step response of the cycle in the presence of a positive step. The response time is not affected by the load. Here, $K_1/X_{\text{tot}} = K_2/X_{\text{tot}} = 0.1$, $k_1 = k_2 = 1$ min^{-1}, and $\lambda = 5$. In the plots, the concentration X^* is normalized by X_{tot}.

which is the same expression for the isolated system in which X^* is scaled by $(1 + \lambda)$. So, the rise time is not affected. The response of the cycle to positive and negative step changes of the input stimulus Z_{tot} are shown in Figure 6.8.

In order to understand how the bandwidth of the system is affected by retroactivity, we consider $Z_{\text{tot}}(t) = \bar{Z} + A_0 \sin(\omega t)$. Let X_e^* be the equilibrium of X^* corresponding to $\bar{Z}$. Let $z = Z_{\text{tot}} - \bar{Z}$ and $x = X^* - X_e^*$ denote small perturbations about the equilibrium. The linearization of system (6.9) is given by

$$\frac{dx}{dt} = -a(\lambda)x + b(\lambda)z(t),$$

in which

$$a(\lambda) = \frac{1}{1 + \lambda}\left(k_1 \bar{Z}\frac{K_1(1 + \lambda)}{((X_{\text{tot}} - X_e^*(1 + \lambda)) + K_1)^2} + k_2 Y_{\text{tot}}\frac{K_2}{(X_e^* + K_2)^2}\right)$$

and

$$b(\lambda) = \frac{k_1}{1 + \lambda}\left(\frac{X_{\text{tot}} - X_e^*(1 + \lambda)}{(X_{\text{tot}} - X_e^*(1 + \lambda)) + K_1}\right),$$

so that the bandwidth of the system is given by $\omega_B = a(\lambda)$.

Figure 6.9 shows the behavior of the bandwidth as a function of the load. When the isolated system steady state input/output characteristic curves are linear-like ($K_1, K_2 \gg X_{\text{tot}}$), the bandwidth monotonically decreases with the load. By contrast, when the isolated system static characteristics are ultrasensitive ($K_1, K_2 \ll X_{\text{tot}}$), the bandwidth of the connected system can be larger than that of the isolated system for sufficiently large amounts of loads. In these conditions, one should expect that

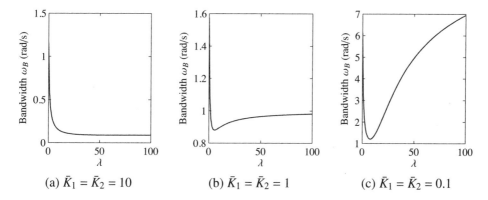

(a) $\bar{K}_1 = \bar{K}_2 = 10$ (b) $\bar{K}_1 = \bar{K}_2 = 1$ (c) $\bar{K}_1 = \bar{K}_2 = 0.1$

Figure 6.9: Behavior of the bandwidth as a function of the effective load λ for different values of the constants $\bar{K}_1, \bar{K}_2$.

the response of the connected system becomes faster than that of the isolated system. These theoretical predictions have been experimentally validated in a covalent modification cycle reconstituted in vitro [52].

6.5 Insulation devices: Retroactivity attenuation

As explained in the previous section, it is not always possible or advantageous to design the downstream system, which we here call module B, such that it applies low retroactivity to the upstream system, here called module A. In fact, module B may already have been designed and optimized for other purposes. A different approach, in analogy to what is performed in electrical circuits, is to design a device to be placed between module A and module B (Figure 6.10) such that the device can transmit the output signal of module A to module B even in the presence of large retroactivity s. That is, the output y of the device should follow the behavior of the output of module A independent of a potentially large load applied by module B. This way module B will receive the desired input signal.

Specifically, consider a system S such as the one shown in Figure 6.3. We would

Figure 6.10: An insulation device is placed between an upstream module A and a downstream module B in order to protect these systems from retroactivity. An insulation device should have $r \approx 0$ and the dynamic response of y to u should be practically independent of s.

like to design such a system such that

(i) the retroactivity r to the input is very small;

(ii) the effect of the retroactivity s on the system is very small (retroactivity attenuation); and

(iii) when $s = 0$, we have that $y \approx Ku$ for some $K > 0$.

Such a system is said to have the *insulation* property and will be called an insulation device. Indeed, such a system does not affect an upstream system because $r \approx 0$ (requirement (i)), it keeps the same output signal $y(t)$ *independently* of any connected downstream system (requirement (ii)), and the output is a linear function of the input in the absence of retroactivity to the output (requirement (iii)). This requirement rules out trivial cases in which y is saturated to a maximal level for all values of the input, leading to no signal transmission. Of course, other requirements may be important, such as the stability of the device and especially the speed of response.

Equation (6.6) quantifies the effect of retroactivity on the dynamics of X as a function of biochemical parameters. These parameters are the affinity of the binding site $1/K_d$, the total concentration of such binding site p_{tot}, and the level of the signal $X(t)$. Therefore, to reduce retroactivity, we can choose parameters such that $\mathcal{R}(X)$ in equation (6.6) is small. A sufficient condition is to choose K_d large (low affinity) and p_{tot} small, for example. Having a small value of p_{tot} and/or low affinity implies that there is a small "flow" of protein X toward its target sites. Thus, we can say that a low retroactivity to the input is obtained when the "input flow" to the system is small. In the next sections, we focus on the retroactivity to the output, that is, on the retroactivity attenuation problem, and illustrate how the problem of designing a device that is robust to s can be formulated as a classical disturbance attenuation problem (Section 3.2). We provide two main design techniques to attenuate retroactivity: the first one is based on the idea of high gain feedback, while the second one uses time scale separation and leverages the structure of the interconnection.

Attenuation of retroactivity to the output using high gain feedback

The basic mechanism for retroactivity attenuation is based on the concept of disturbance attenuation through high gain feedback presented in Section 3.2. In its simplest form, it can be illustrated by the diagram of Figure 6.11a, in which the retroactivity to the output s plays the same role as an additive disturbance. For large gains G, the effect of the retroactivity s to the output is negligible as the following simple computation shows. The output y is given by

$$y = G(u - Ky) + s,$$

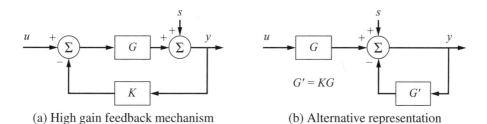

(a) High gain feedback mechanism   (b) Alternative representation

Figure 6.11: The block diagram in (a) shows the basic high gain feedback mechanism to attenuate the contribution of disturbance s to the output y. The diagram in (b) shows an alternative representation, which will be employed to design biological insulation devices.

which leads to

$$y = u \frac{G}{1 + KG} + \frac{s}{1 + KG}.$$

As G grows, y tends to u/K, which is independent of the retroactivity s.

Figure 6.11b illustrates an alternative representation of the diagram depicting high gain feedback. This alternative depiction is particularly useful as it highlights that to attenuate retroactivity we need to (1) amplify the input of the system through a large gain and (2) apply a similarly large negative feedback on the output. The question of how to realize a large input amplification and a similarly large negative feedback on the output through biomolecular interactions is the subject of the next section. In what follows, we first illustrate how this strategy also works for a dynamical system of the form of equation (6.5).

Consider the dynamics of the connected transcriptional system given by

$$\frac{dX}{dt} = (k(t) - \gamma X)\left(\frac{1}{1 + \mathcal{R}(X)}\right).$$

Assume that we can apply a gain G to the input $k(t)$ and a negative feedback gain G' to X with $G' = KG$. This leads to the new differential equation for the connected system given by

$$\frac{dX}{dt} = (Gk(t) - (G' + \gamma)X)(1 - d(t)), \tag{6.10}$$

in which we have defined $d(t) = \mathcal{R}(X)/(1 + \mathcal{R}(X))$. Since $d(t) < 1$, we can verify (see Exercise 6.8) that as G grows $X(t)$ tends to $k(t)/K$ for both the connected system (6.10) and the isolated system

$$\frac{dX}{dt} = Gk(t) - (G' + \gamma)X. \tag{6.11}$$

Specifically, we have the following fact:

Proposition 6.1. *Consider the scalar system* $\dot{x} = G(t)(k(t) - Kx)$ *with* $G(t) \geq G_0 > 0$ *and* $\dot{k}(t)$ *bounded. Then, there are positive constants* C_0 *and* C_1 *such that*

$$\left| x(t) - \frac{k(t)}{K} \right| \leq C_0 e^{-G_0 Kt} + \frac{C_1}{G_0}.$$

To derive this result, we can explicitly integrate the system since it is linear (time-varying). For details, the reader is referred to [22]. The solutions $X(t)$ of the connected and isolated systems thus tend to each other as G increases, implying that the presence of the disturbance $d(t)$ will not significantly affect the time behavior of $X(t)$. It follows that the effect of retroactivity can be arbitrarily attenuated by increasing gains G and G'.

The next question we address is how we can implement such amplification and feedback gains in a biomolecular system.

Biomolecular realizations of high gain feedback

In this section, we illustrate two possible biomolecular implementations to obtain a large input amplification gain and a similarly large negative feedback on the output. Both implementations realize the negative feedback through enhanced degradation. The first design realizes amplification through transcriptional activation, while the second design uses phosphorylation.

Design 1: Amplification through transcriptional activation

This design is depicted in Figure 6.12. We implement a large amplification of the input signal $Z(t)$ by having Z be a transcriptional activator for protein X, such that the promoter p_0 controlling the expression of X is a strong, non-leaky promoter activated by Z. The signal $Z(t)$ can be further amplified by increasing the strength of the ribosome binding site of gene x. The negative feedback mechanism on X relies on enhanced degradation of X. Since this must be large, one possible way to obtain an enhanced degradation for X is to have a specific protease, called Y, be expressed by a strong constitutive promoter.

To investigate whether such a design realizes a large amplification and a large negative feedback on X as needed to attenuate retroactivity to the output, we construct a model. The reaction of the protease Y with protein X is modeled as the two-step reaction

$$X + Y \underset{d}{\overset{a}{\rightleftharpoons}} W \overset{\bar{k}}{\rightarrow} Y.$$

The input/output system model of the insulation device that takes Z as an input and

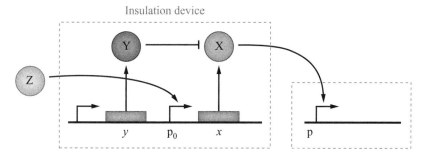

Figure 6.12: Implementation of high gain feedback (Design 1). The input $Z(t)$ is amplified by virtue of a strong promoter p_0. The negative feedback on the output X is obtained by enhancing its degradation through the protease Y.

gives X as an output is given by the following equations:

$$\frac{dZ}{dt} = k(t) - \gamma_Z Z + \left[k'_{\text{off}} \bar{C} - k'_{\text{on}} Z(p_{0,\text{tot}} - \bar{C}) \right], \tag{6.12}$$

$$\frac{d\bar{C}}{dt} = k'_{\text{on}} Z(p_{0,\text{tot}} - \bar{C}) - k'_{\text{off}} \bar{C}, \tag{6.13}$$

$$\frac{dm_X}{dt} = G\bar{C} - \delta m_X, \tag{6.14}$$

$$\frac{dW}{dt} = aXY - dW - \bar{k}W, \tag{6.15}$$

$$\frac{dY}{dt} = -aYX + \bar{k}W + \alpha G - \gamma_Y Y + dW, \tag{6.16}$$

$$\frac{dX}{dt} = \kappa m_X - aYX + dW - \gamma_X X + \left[k_{\text{off}} C - k_{\text{on}} X(p_{\text{tot}} - C) \right], \tag{6.17}$$

$$\frac{dC}{dt} = -k_{\text{off}} C + k_{\text{on}} X(p_{\text{tot}} - C), \tag{6.18}$$

in which we have assumed that the expression of gene z is controlled by a promoter with activity $k(t)$. In this system, we have denoted by k'_{on} and k'_{off} the association and dissociation rate constants of Z with its promoter site p_0 in total concentration $p_{0,\text{tot}}$. Also, $\bar{C}$ is the complex of Z with such a promoter site. Here, m_X is the mRNA of X, and C is the complex of X bound to the downstream binding sites p with total concentration p_{tot}. The promoter controlling gene y has strength αG, for some constant α, and it has about the same strength as the promoter controlling gene x.

The terms in the square brackets in equation (6.12) represent the retroactivity r to the input of the insulation device in Figure 6.12. The terms in the square brackets in equation (6.17) represent the retroactivity s to the output of the insulation device. The dynamics of equations (6.12)–(6.18) without s describe the dynamics of X with no downstream system (isolated system).

Equations (6.12) and (6.13) determine the signal $\bar{C}(t)$ that is the input to equations (6.14)–(6.18). For the discussion regarding the attenuation of the effect of s, it is not relevant what the specific form of signal $\bar{C}(t)$ is. Let then $\bar{C}(t)$ be any bounded signal. Since equation (6.14) takes $\bar{C}(t)$ as an input, we will have that $m_X(t) = Gv(t)$, for a suitable signal $v(t)$. Let us assume for the sake of simplifying the analysis that the protease reaction is a one-step reaction. Therefore, equation (6.16) simplifies to

$$\frac{dY}{dt} = \alpha G - \gamma_Y Y$$

and equation (6.17) simplifies to

$$\frac{dX}{dt} = \kappa m_X - \bar{k}'YX - \gamma_X X + k_{\text{off}}C - k_{\text{on}}X(p_{\text{tot}} - C),$$

for a suitable positive constant $\bar{k}'$. If we further consider the protease to be at its equilibrium, we have that $Y(t) = \alpha G/\gamma_Y$.

As a consequence, the X dynamics become

$$\frac{dX}{dt} = \kappa Gv(t) - (\bar{k}'\alpha G/\gamma_Y + \gamma_X)X + k_{\text{off}}C - k_{\text{on}}X(p_{\text{tot}} - C),$$

with C determined by equation (6.18). By using the same singular perturbation argument employed in the previous section, the dynamics of X can be reduced to

$$\frac{dX}{dt} = (\kappa Gv(t) - (\bar{k}'\alpha G/\gamma_Y + \gamma_X)X)(1 - d(t)), \tag{6.19}$$

in which $0 < d(t) < 1$ is the retroactivity term given by $\mathcal{R}(X)/(1 + \mathcal{R}(X))$. Then, as G increases, $X(t)$ becomes closer to the solution of the isolated system

$$\frac{dX}{dt} = \kappa Gv(t) - (\bar{k}'\alpha G/\gamma_Y + \gamma_X)X,$$

as explained in the previous section by virtue of Proposition 6.1.

We now turn to the question of minimizing the retroactivity to the input r because its effect can alter the input signal $Z(t)$. In order to decrease r, we must guarantee that the retroactivity measure given in equation (6.6), in which we substitute Z in place of X, $p_{0,\text{tot}}$ in place of p_{tot}, and $K_d' = k_{\text{on}}'/k_{\text{off}}'$ in place of K_d, is small. This is the case if $K_d' \gg Z$ and $p_{0,\text{tot}}/K_d' \ll 1$.

Simulation results for the system described by equations (6.12)–(6.18) are shown in Figure 6.13. For large gains ($G = 1000$, $G = 100$), the performance considerably improves compared to the case in which X was generated by a transcriptional component accepting Z as an input (Figure 6.5). For lower gains ($G = 10$, $G = 1$), the performance starts to degrade and becomes poor for $G = 1$. Since we can view G as the number of transcripts produced per unit of time (one minute) per complex of protein Z bound to promoter p_0, values $G = 100, 1000$ may be difficult to realize in vivo, while the values $G = 10, 1$ could be more easily realized.

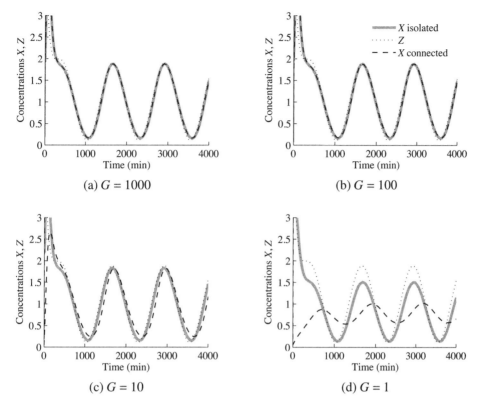

Figure 6.13: Results for different gains G (Design 1). In all plots, $k(t) = 0.01(1 + sin(\omega t))$, $p_{tot} = 100$ nM, $k_{off} = 10$ min^{-1}, $k_{on} = 10$ min^{-1} nM^{-1}, $\gamma_Z = 0.01 = \gamma_Y$ min^{-1}, and $\omega = 0.005$ rad/min. Also, we have set $\delta = 0.01$ min^{-1}, $p_{0,tot} = 1$ nM, $a = 0.01$ min^{-1} nM^{-1}, $d = \bar{k}' = 0.01$ min^{-1}, $k'_{off} = 200$ min^{-1}, $k'_{on} = 10$ min^{-1} nM^{-1}, $\alpha = 0.1$ nM/min, $\gamma_X = 0.1$ min^{-1}, $\kappa = 0.1$ min^{-1}, and $G = 1000, 100, 10, 1$. The retroactivity to the output is not well attenuated for $G = 1$ and the attenuation capability begins to worsen for $G = 10$.

However, the value of κ increases with the strength of the ribosome binding site and therefore the gain may be further increased by picking strong ribosme binding sites for x. The values of the parameters chosen in Figure 6.13 are such that $K'_d \gg Z$ and $p_{0,tot} \ll K'_d$. This is enough to guarantee that there is small retroactivity r to the input of the insulation device. The poorer performance of the device for $G = 1$ is therefore entirely due to poor attenuation of the retroactivity s to the output. To obtain a large negative feedback gain, we also require high expression of the protease. It is therefore important that the protease is highly specific to its target X.

Design 2: Amplification through phosphorylation

In this design, the amplification gain G of Z is obtained by having Z be a kinase that phosphorylates a substrate X, which is available in abundance. The negative

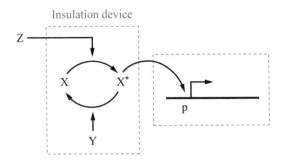

Figure 6.14: Implementation of high gain feedback (Design 2). Amplification of Z occurs through the phosphorylation of substrate X. Negative feedback occurs through a phosphatase Y that converts the active form X* back to its inactive form X.

feedback gain G' on the phosphorylated protein X^* is obtained by having a phosphatase Y dephosphorylate the active protein X^*. Protein Y should also be available in abundance in the system. This implementation is depicted in Figure 6.14.

To illustrate what key parameters enable retroactivity attenuation, we first consider a simplified model for the phosphorylation and dephosphorylation processes. This model will help in obtaining a conceptual understanding of what reactions are responsible in realizing the desired gains G and G'. The one-step model that we consider is the same as considered in Chapter 2 (Exercise 2.12):

$$Z + X \xrightarrow{k_1} Z + X^*, \qquad Y + X^* \xrightarrow{k_2} Y + X.$$

We assume that there is an abundance of protein X and of phosphatase Y in the system and that these quantities are conserved. The conservation of X gives $X + X^* + C = X_{\text{tot}}$, in which X is the inactive protein, X^* is the phosphorylated protein that binds to the downstream sites p, and C is the complex of the phosphorylated protein X^* bound to the promoter p. The X^* dynamics can be described by the following model:

$$\frac{dX^*}{dt} = k_1 X_{\text{tot}} Z(t) \left(1 - \frac{X^*}{X_{\text{tot}}} - \left[\frac{C}{X_{\text{tot}}} \right] \right) - k_2 Y X^* + [k_{\text{off}} C - k_{\text{on}} X^* (p_{\text{tot}} - C)],$$

$$\frac{dC}{dt} = -k_{\text{off}} C + k_{\text{on}} X^* (p_{\text{tot}} - C). \tag{6.20}$$

The two terms in the square brackets represent the retroactivity s to the output of the insulation device of Figure 6.14. For a weakly activated pathway [41], $X^* \ll X_{\text{tot}}$. Also, if we assume that the total concentration of X is large compared to the concentration of the downstream binding sites, that is, $X_{\text{tot}} \gg p_{\text{tot}}$, equation (6.20) is approximatively equal to

$$\frac{dX^*}{dt} = k_1 X_{\text{tot}} Z(t) - k_2 Y X^* + k_{\text{off}} C - k_{\text{on}} X^* (p_{\text{tot}} - C).$$

Let $G = k_1 X_{\text{tot}}$ and $G' = k_2 Y$. Exploiting again the difference of time scales between the X^* dynamics and the C dynamics, the dynamics of X^* can be finally reduced to

$$\frac{dX^*}{dt} = (GZ(t) - G'X^*)(1 - d(t)),$$

in which $0 < d(t) < 1$ is the retroactivity term. Therefore, for G and G' large enough, $X^*(t)$ tends to the solution $X^*(t)$ of the isolated system

$$\frac{dX^*}{dt} = GZ(t) - G'X^*,$$

as explained before by virtue of Proposition 6.1. It follows that the effect of the retroactivity to the output s is attenuated by increasing the effective rates $k_1 X_{\text{tot}}$ and $k_2 Y$. That is, to obtain large input and negative feedback gains, one should have large phosphorylation/dephosphorylation rates and/or a large amount of protein X and phosphatase Y in the system. This reveals that the values of the phosphorylation/dephosphorylation rates cover an important role toward the retroactivity attenuation property of the insulation device of Figure 6.14. From a practical point of view, the effective rates can be increased by increasing the total amounts of X and Y. These amounts can be tuned, for example, by placing the x and y genes under the control of inducible promoters. The reader can verify through simulation how the effect of retroactivity can be attenuated by increasing the phosphatase and substrate amounts (see Exercise 6.9). Experiments performed on a covalent modification cycle reconstituted in vitro confirmed that increasing the effective rates of modification is an effective means to attain retroactivity attenuation [52].

A design similar to the one illustrated here can be proposed in which a phosphorylation cascade, such as the MAPK cascade, realizes the input amplification and an explicit feedback loop is added from the product of the cascade to its input [84]. The design presented here is simpler as it involves only one phosphorylation cycle and does not require any explicit feedback loop. In fact, a strong negative feedback can be realized by the action of the phosphatase that converts the active protein form X^* back to its inactive form X.

Attenuation of retroactivity to the output using time scale separation

In this section, we present a more general mechanism for retroactivity attenuation, which can be applied to systems of differential equations of arbitrary dimension. This will allow us to consider more complex and realistic models of the phosphorylation reactions as well as more complicated systems.

For this purpose, consider Figure 6.15. We illustrate next how system S can attenuate retroactivity s by employing time scale separation. Specifically, when the internal dynamics of the system are much faster compared to those of the input u, the system immediately reaches its quasi-steady state with respect to the input. Any

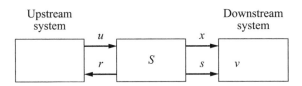

Figure 6.15: Interconnection of a device with input u and output x to a downstream system with internal state v applying retroactivity s.

load-induced delays occur at the faster time scale of system S and thus are negligible in the slower time scale of the input. Therefore, as long as the quasi-steady state is independent of retroactivity s, the system has the retroactivity attenuation property. We show here that the quasi-steady state can be made basically independent of s by virtue of the interconnection structure between the systems.

To illustrate this idea mathematically, consider the following simple structure in which (for simplicity) we assume that all variables are scalar:

$$\frac{du}{dt} = f_0(u,t) + r(u,x), \qquad \frac{dx}{dt} = Gf_1(x,u) + \bar{G}s(x,v), \qquad \frac{dv}{dt} = -\bar{G}s(x,v). \quad (6.21)$$

Here let $G \gg 1$ model the fact that the internal dynamics of the system are much faster than that of the input. Similarly, $\bar{G} \gg 1$ models the fact that the dynamics of the interconnection with downstream systems are also very fast. This is usually the case since the reactions contributing to s are binding/unbinding reactions, which are much faster than most other biochemical processes, including gene expression and phosphorylation. We make the following informal claim:

> If $G \gg 1$ and the Jacobian $\partial f_1(x,u)/\partial x$ has all eigenvalues with negative real part, then $x(t)$ is not affected by retroactivity s after a short initial transient, independently of the value of $\bar{G}$.

A formal statement of this result can be found in [50]. This result states that independently of the characteristics of the downstream system, system S can be tuned (by making G large enough) such that it attenuates the retroactivity to the output. To clarify why this would be the case, it is useful to rewrite system (6.21) in standard singular perturbation form by employing $\epsilon := 1/G$ as a small parameter and $\tilde{x} := x + v$ as the slow variable. Hence, the dynamics can be rewritten as

$$\frac{du}{dt} = f_0(u,t) + r(u,x), \qquad \epsilon \frac{d\tilde{x}}{dt} = f_1(\tilde{x} - v, u), \qquad \frac{dv}{dt} = -\bar{G}s(\tilde{x} - v, v). \quad (6.22)$$

Since $\partial f_1/\partial \tilde{x}$ has eigenvalues with negative real part, one can apply standard singular perturbation to show that after a very fast transient, the trajectories are attracted to the slow manifold given by $f_1(\tilde{x} - v, u) = 0$. This is locally given by $x = g(u)$ solving $f_1(x,u) = 0$. Hence, on the slow manifold we have that $x(t) = g(u(t))$, which is independent of the downstream system, that is, it is not affected by retroactivity.

The same result holds for a more general class of systems in which the variables u, x, v are vectors:

$$\frac{du}{dt} = f_0(u, t) + r(u, x), \qquad \frac{dx}{dt} = Gf_1(x, u) + \bar{G}As(x, v), \qquad \frac{dv}{dt} = -\bar{G}Bs(x, v),$$
(6.23)

as long as there are matrices T and M such that $TA - MB = 0$ and T is invertible. In fact, one can take the system to new coordinates $u, \tilde{x}, v$ with $\tilde{x} = Tx + Mv$, in which the system will have the singular perturbation form (6.22), where the state variables are vectors. Note that matrices A and B are stoichiometry matrices and s represents a vector of reactions, usually modeling binding and unbinding processes. The existence of T and M such that $TA - MB = 0$ models the fact that in these binding reactions species do not get destroyed or created, but simply transformed between species that belong to the upstream system and species that belong to the downstream system.

Biomolecular realizations of time scale separation

We next consider possible biomolecular structures that realize the time scale separation required for insulation. Since this principle is based on a fast time scale of the device dynamics when compared to that of the device input, we focus on signaling systems, which are known to evolve on faster time scales than those of protein production and decay.

Design 1: Implementation through phosphorylation

We consider now a more realistic model for the phosphorylation and dephosphorylation reactions in a phosphorylation cycle than those considered in Section 6.5. In particular, we consider a two-step reaction model as seen in Section 2.4. According to this model, we have the following two reactions for phosphorylation and dephosphorylation:

$$Z + X \underset{d_1}{\overset{a_1}{\rightleftharpoons}} C_1 \overset{k_1}{\longrightarrow} X^* + Z, \qquad Y + X^* \underset{d_2}{\overset{a_2}{\rightleftharpoons}} C_2 \overset{k_2}{\longrightarrow} X + Y.$$
(6.24)

Additionally, we have the conservation equations $Y_{\text{tot}} = Y + C_2$, $X_{\text{tot}} = X + X^* + C_1 + C_2 + C$, because proteins X and Y are not degraded. Therefore, the differential equations modeling the system of Figure 6.14 become

$$\frac{dZ}{dt} = k(t) - \gamma Z \left[-a_1 Z X_{\text{tot}} \left(1 - \frac{X^*}{X_{\text{tot}}} - \frac{C_1}{X_{\text{tot}}} - \frac{C_2}{X_{\text{tot}}} - \left[\frac{C}{X_{\text{tot}}} \right] \right) + (d_1 + k_1) C_1 \right], \quad (6.25)$$

$$\frac{dC_1}{dt} = -(d_1 + k_1) C_1 + a_1 Z X_{\text{tot}} \left(1 - \frac{X^*}{X_{\text{tot}}} - \frac{C_1}{X_{\text{tot}}} - \frac{C_2}{X_{\text{tot}}} - \left[\frac{C}{X_{\text{tot}}} \right] \right), \quad (6.26)$$

$$\frac{dC_2}{dt} = -(k_2 + d_2)C_2 + a_2 Y_{\text{tot}} X^* \left(1 - \frac{C_2}{Y_{\text{tot}}}\right), \tag{6.27}$$

$$\frac{dX^*}{dt} = k_1 C_1 + d_2 C_2 - a_2 Y_{\text{tot}} X^* \left(1 - \frac{C_2}{Y_{\text{tot}}}\right) + \left[k_{\text{off}} C - k_{\text{on}} X^* (p_{\text{tot}} - C)\right], \tag{6.28}$$

$$\frac{dC}{dt} = -k_{\text{off}} C + k_{\text{on}} X^* (p_{\text{tot}} - C), \tag{6.29}$$

in which the production of Z is controlled by a promoter with activity $k(t)$. The terms in the large square bracket in equation (6.25) represent the retroactivity r to the input, while the terms in the square brackets of equations (6.26) and (6.28) represent the retroactivity s to the output.

We assume that $X_{\text{tot}} \gg p_{\text{tot}}$ so that in equations (6.25) and (6.26) we can neglect the term C/X_{tot} since $C < p_{\text{tot}}$. Choose X_{tot} to be sufficiently large so that $G = a_1 X_{\text{tot}}/\gamma \gg 1$. Also, let $\bar{G} = k_{\text{off}}/\gamma$, which is also much larger than 1 since binding reactions are much faster than protein production and decay rates ($k_{\text{off}} \gg \gamma$) and write $k_{\text{on}} = k_{\text{off}}/K_d$. Choosing Y_{tot} to also be sufficiently large, we can guarantee that $a_2 Y_{\text{tot}}$ is of the same order as $a_1 X_{\text{tot}}$ and we can let $\alpha_1 = a_1 X_{\text{tot}}/(\gamma G)$, $\alpha_2 = a_2 Y_{\text{tot}}/(\gamma G)$, $\delta_1 = d_1/(\gamma G)$, and $\delta_2 = d_2/(\gamma G)$. Finally, since the catalytic rate constants k_1, k_2 are much larger than protein decay, we can assume that they are of the same order of magnitude as $a_1 X_{\text{tot}}$ and $a_2 Y_{\text{tot}}$, so that we define $c_i = k_i/(\gamma G)$. With these, letting $z = Z + C_1$ we obtain the system in the form

$$\frac{dz}{dt} = k(t) - \gamma(z - C_1),$$

$$\frac{dC_1}{dt} = G\left(-\gamma(\delta_1 + c_1)C_1 + \gamma\alpha_1(z - C_1)\left(1 - \frac{X^*}{X_{\text{tot}}} - \frac{C_1}{X_{\text{tot}}} - \frac{C_2}{X_{\text{tot}}}\right)\right),$$

$$\frac{dC_2}{dt} = G\left(-\gamma(\delta_2 + c_2)C_2 + \gamma\alpha_2 X^*\left(1 - \frac{C_2}{Y_{\text{tot}}}\right)\right), \tag{6.30}$$

$$\frac{dX^*}{dt} = G\left(\gamma c_1 C_1 + \gamma\delta_2 C_2 - \gamma\alpha_2 X^*\left(1 - \frac{C_2}{Y_{\text{tot}}}\right)\right) + \bar{G}\left(\gamma C - \gamma/K_d(p_{\text{tot}} - C)X^*\right),$$

$$\frac{dC}{dt} = -\bar{G}\left(\gamma C - \gamma/K_d(p_{\text{tot}} - C)X^*\right),$$

which is in the form of system (6.23) with $u = z$, $x = (C_1, C_2, X^*)$, and $v = C$, in which we can choose T as the 3×3 identity matrix and

$$M = \begin{pmatrix} 0 \\ 0 \\ 1 \end{pmatrix}.$$

It is also possible to show that the Jacobian of f_1 has all eigenvalues with negative real part (see Exercise 6.11). Hence, for G sufficiently larger than one, this system attenuates the effect of the retroactivity to the output s. For G to be large, one has to require that $a_1 X_{\text{tot}}$ is sufficiently large and that $a_2 Y_{\text{tot}}$ is also comparatively large.

These are compatible with the design requirements obtained in the previous section based on the one-step reaction model of the enzymatic reactions.

In order to understand the effect of retroactivity to the input on the Z dynamics, one can consider the reduced system describing the dynamics on the time scale of Z. To this end, let $K_{m,1} = (d_1 + k_1)/a_1$ and $K_{m,2} = (d_2 + k_2)/a_2$ represent the Michaelis-Menten constants of the forward and backward enzymatic reactions, let $G = 1/\epsilon$ in equations (6.30), and take ϵ to the left-hand side. Setting $\epsilon = 0$, the following relationships can be obtained:

$$C_1 = g_1(X^*) = \frac{(X^* Y_{\text{tot}} k_2)/(K_{m,2} k_1)}{1 + X^*/K_{m,2}}, \qquad C_2 = g_2(X^*) = \frac{(X^* Y_{\text{tot}})/K_{m,2}}{1 + X^*/K_{m,2}}. \qquad (6.31)$$

Using expressions (6.31) in the second of equations (6.30) with $\epsilon = 0$ leads to

$$g_1(X^*)\left(\delta_1 + c_1 + \frac{\alpha_1 Z}{X_{\text{tot}}}\right) = \alpha_1 Z\left(1 - \frac{X^*}{X_{\text{tot}}} - \frac{g_2(X^*)}{X_{\text{tot}}}\right). \qquad (6.32)$$

Assuming for simplicity that $X^* \ll K_{m,2}$, we obtain that

$$g_1(X^*) \approx (X^* Y_{\text{tot}} k_2)/(K_{m,2} k_1)$$

and that

$$g_2(X^*) \approx X^*/K_{m,2} Y_{\text{tot}}.$$

As a consequence of these simplifications, equation (6.32) leads to

$$X^*(Z) = \frac{\alpha_1 Z}{(\alpha_1 Z/X_{\text{tot}})(1 + Y_{\text{tot}}/K_{m,2} + (Y_{\text{tot}} k_2)/(K_{m,2} k_1)) + (Y_{\text{tot}} k_2)/(K_{m,2} k_1)(\delta_1 + c_1)}.$$

In order not to have distortion from Z to X^*, we require that

$$Z \ll \frac{Y_{\text{tot}}(k_2/k_1)(K_m/K_{m,2})}{1 + Y_{\text{tot}}/K_{m,2} + (Y_{\text{tot}}/K_{m,2})(k_2/k_1)}, \qquad (6.33)$$

so that $X^*(Z) \approx Z(X_{\text{tot}} K_{m,2} k_1)/(Y_{\text{tot}} K_{m,1} k_2)$ and therefore we have a linear relationship between X^* and Z with gain from Z to X^* given by $(X_{\text{tot}} K_{m,2} k_1)/(Y_{\text{tot}} K_{m,1} k_2)$. In order not to have attenuation from Z to X^* we require that the gain is greater than or equal to one, that is,

$$\text{input/output gain} \approx \frac{X_{\text{tot}} K_{m,2} k_1}{Y_{\text{tot}} K_{m,1} k_2} \geq 1. \qquad (6.34)$$

Requirements (6.33), (6.34) and $X^* \ll K_{m,2}$ are enough to guarantee that we do not have nonlinear distortion between Z and X^* and that X^* is not attenuated with respect to Z. In order to guarantee that the retroactivity r to the input is sufficiently small, we need to quantify the retroactivity effect on the Z dynamics due to the binding of Z with X. To achieve this, we proceed as in Section 6.3 by computing the Z dynamics on the slow manifold, which gives a good approximation of the dynamics of Z if $\epsilon \approx 0$. These dynamics are given by

$$\frac{dZ}{dt} = (k(t) - \gamma Z)\left(1 - \frac{dg_1}{dX^*}\frac{dX^*}{dz}\right),$$

in which $(dg_1/dX^*)(dX^*/dz)$ measures the effect of the retroactivity r to the input on the Z dynamics. Direct computation of dg_1/dX^* and of dX^*/dz along with $X^* \ll K_{m,2}$ and with (6.33) leads to $(dg_1/dX^*)(dX^*/dz) \approx X_{\text{tot}}/K_{m,1}$, so that in order to have small retroactivity to the input, we require that

$$\frac{X_{\text{tot}}}{K_{m,1}} \ll 1. \tag{6.35}$$

Hence, a design tradeoff appears: X_{tot} should be sufficiently large to provide a gain G large enough to attenuate the retroactivity to the output. Yet, X_{tot} should be small enough compared to $K_{m,1}$ to apply minimal retroactivity to the input.

In conclusion, in order to have attenuation of the effect of the retroactivity to the output s, we require that the time scale of the phosphorylation/dephosphorylation reactions is much faster than the production and decay processes of Z (the input to the insulation device) and that $X_{\text{tot}} \gg p_{\text{tot}}$, that is, the total amount of protein X is in abundance compared to the downstream binding sites p. To also obtain a small effect of the retroactivity to the input, we require that $K_{m,1} \gg X_{\text{tot}}$. This is satisfied if, for example, kinase Z has low affinity to binding with X. To keep the input/output gain between Z and X^* close to one (from equation (6.34)), one can choose $X_{\text{tot}} = Y_{\text{tot}}$, and equal coefficients for the phosphorylation and dephosphorylation reactions, that is, $K_{m,1} = K_{m,2}$ and $k_1 = k_2$.

The system in equations (6.25)–(6.29) was simulated with and without the downstream binding sites p, that is, with and without, respectively, the terms in the small box of equation (6.25) and in the boxes in equations (6.28) and (6.26). This is performed to highlight the effect of the retroactivity to the output s on the dynamics of X^*. The simulations validate our theoretical study that indicates that when $X_{\text{tot}} \gg p_{\text{tot}}$ and the time scales of phosphorylation/dephosphorylation are much faster than the time scale of decay and production of the protein Z, the retroactivity to the output s is very well attenuated (Figure 6.16a). Similarly, the time behavior of Z was simulated with and without the terms in the square brackets in equation (6.25), which represent the retroactivity to the input r, to verify whether the insulation device exhibits small retroactivity to the input r. The similarity of the behaviors of $Z(t)$ with and without its downstream binding sites on X (Figure 6.16a) indicates that there is no substantial retroactivity to the input r generated by the insulation device. This is obtained because $X_{\text{tot}} \ll K_{m,1}$ as indicated in equation (6.35), in which $1/K_m$ can be interpreted as the affinity of the binding of X to Z.

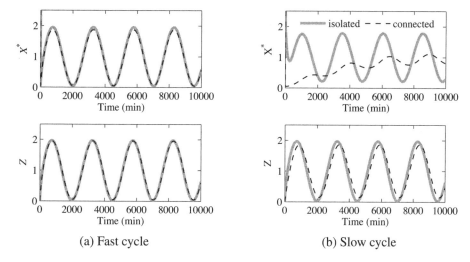

(a) Fast cycle   (b) Slow cycle

Figure 6.16: Time scale separation mechanism for insulation: Implementation through phosphorylation. Simulation results for system in equations (6.25)–(6.29). In all plots, $p_{tot} = 100$ nM, $k_{off} = 10$ min^{-1}, $k_{on} = 10$ min^{-1} nM^{-1}, $\gamma = 0.01$ min^{-1}, $k(t) = 0.01(1 + sin(\omega t))$ min^{-1}, and $\omega = 0.005$ rad/min. (a) Performance with fast phosphorylation cycle. Here, $k_1 = k_2 = 50$ min^{-1}, $a_2 = a_1 = 0.01$ min^{-1} nM^{-1}, $d_1 = d_2 = 10$ min^{-1}, and $Y_{tot} = X_{tot} = 1500$ nM. The small error shows that the effect of the retroactivity to the output s is attenuated very well. In the Z plot, the isolated system stands for the case in which Z does not have X to bind to, while the connected system stands for the case in which Z binds to substrate X. The small error confirms a small retroactivity to the input r. (b) Performance with a slow phosphorylation cycle. Here, we set $k_1 = k_2 = 0.01$ min^{-1}, while the other parameters are left the same.

Our simulation study also indicates that a faster time scale of the phosphorylation/dephosphorylation reactions is necessary, even for high values of X_{tot} and Y_{tot}, to maintain perfect attenuation of the retroactivity to the output s and small retroactivity to the output r. In fact, slowing down the time scale of phosphorylation and dephosphorylation, the system loses its insulation property (Figure 6.16b). In particular, the attenuation of the effect of the retroactivity to the output s is lost because there is not enough separation of time scales between the Z dynamics and the internal device dynamics. The device also displays a non-negligible amount of retroactivity to the input because the condition $K_m \ll X_{tot}$ is not satisfied anymore.

Design 2: Realization through phosphotransfer

Here we illustrate that another possible implementation of the mechanism for insulation based on time scale separation is provided by phosphotransfer systems. These systems, just like phosphorylation cycles, have a very fast dynamics when compared to gene expression. Specifically, we consider the realization shown in Figure 6.17, in which the input is a phosphate donor Z and the output is the active

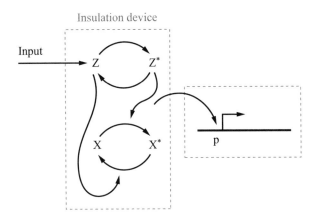

Figure 6.17: The insulation device is a phosphotransfer system. The output X^* activates transcription through the reversible binding of X^* to downstream DNA promoter sites p.

transcription factor X^*. We let X be the transcription factor in its inactive form and let Z^* be the active phosphate donor, that is, a protein that can transfer its phosphate group to the acceptor X. The standard phosphotransfer reactions can be modeled according to the two-step reaction model

$$Z^* + X \underset{k_2}{\overset{k_1}{\rightleftharpoons}} C_1 \underset{k_4}{\overset{k_3}{\rightleftharpoons}} X^* + Z,$$

in which C_1 is the complex of Z bound to X bound to the phosphate group. Additionally, we assume that protein Z can be phosphorylated and protein X^* dephosphorylated by other phosphotransfer interactions. These reactions are modeled as one-step reactions depending only on the concentrations of Z and X^*, that is,

$$Z \overset{\pi_1}{\rightarrow} Z^*, \qquad X^* \overset{\pi_2}{\rightarrow} X.$$

Protein X is assumed to be conserved in the system, that is, $X_{tot} = X + C_1 + X^* + C$. We assume that protein Z is produced with time-varying production rate $k(t)$ and decays with rate γ. The active transcription factor X^* binds to downstream DNA binding sites p with total concentration p_{tot} to activate transcription through the reversible reaction

$$X^* + p \underset{k_{off}}{\overset{k_{on}}{\rightleftharpoons}} C.$$

Since the total amount of p is conserved, we also have that $C + p = p_{tot}$. The ODE model corresponding to this system is thus given by the equations

$$\frac{dZ}{dt} = k(t) - \gamma Z + k_3 C_1 - k_4 X^* Z - \pi_1 Z,$$

$$\frac{dC_1}{dt} = k_1 X_{tot} \left(1 - \frac{X^*}{X_{tot}} - \frac{C_1}{X_{tot}} - \left[\frac{C}{X_{tot}} \right] \right) Z^* - k_3 C_1 - k_2 C_1 + k_4 X^* Z,$$

$$\frac{dZ^*}{dt} = \pi_1 Z + k_2 C_1 - k_1 X_{\text{tot}} \left(1 - \frac{X^*}{X_{\text{tot}}} - \frac{C_1}{X_{\text{tot}}} - \left[\frac{C}{X_{\text{tot}}} \right] \right) Z^*,$$

$$\frac{dX^*}{dt} = k_3 C_1 - k_4 X^* Z + [k_{\text{off}} C - k_{\text{on}} X^* (p_{\text{tot}} - C)] - \pi_2 X^*,$$

$$\frac{dC}{dt} = k_{\text{on}} X^* (p_{\text{tot}} - C) - k_{\text{off}} C.$$

Just like phosphorylation, phosphotransfer reactions are much faster than protein production and decay. Hence, as performed before, define $G = X_{\text{tot}} k_1 / \gamma$ so that $\bar{k}_1 = X_{\text{tot}} k_1 / G$, $\bar{k}_2 = k_2 / G$, $\bar{k}_3 = k_3 / G$, $\bar{k}_4 = X_{\text{tot}} k_4 / G$, $\bar{\pi}_1 = \pi_1 / G$, and $\bar{\pi}_2 = \pi_2 / G$ are of the same order of $k(t)$ and γ. Similarly, the process of protein binding and unbinding to promoter sites is much faster than protein production and decay. We let $\bar{G} = k_{\text{off}} / \gamma$ and $K_d = k_{\text{off}} / k_{\text{on}}$. Assuming also that $p_{\text{tot}} \ll X_{\text{tot}}$, we have that $C \ll X_{\text{tot}}$ so that the above system can be rewritten as

$$\frac{dZ}{dt} = k(t) - \gamma Z - G \bar{\pi}_1 Z + G \left(\bar{k}_3 C_1 - \bar{k}_4 \left(\frac{X^*}{X_{\text{tot}}} \right) Z \right),$$

$$\frac{dC_1}{dt} = G \left(\bar{k}_1 \left(1 - \frac{X^*}{X_{\text{tot}}} - \frac{C_1}{X_{\text{tot}}} \right) Z^* - \bar{k}_3 C_1 - \bar{k}_2 C_1 + \bar{k}_4 \left(\frac{X^*}{X_{\text{tot}}} \right) Z \right),$$

$$\frac{dZ^*}{dt} = G \left(\bar{\pi}_1 Z + \bar{k}_2 C_1 - \bar{k}_1 \left(1 - \frac{X^*}{X_{\text{tot}}} - \frac{C_1}{X_{\text{tot}}} \right) Z^* \right), \qquad (6.36)$$

$$\frac{dX^*}{dt} = G \left(\bar{k}_3 C_1 - \bar{k}_4 \left(\frac{X^*}{X_{\text{tot}}} \right) Z - \bar{\pi}_2 X^* \right) + \bar{G} \left(\gamma C - \frac{\gamma}{K_d} X^* (p_{\text{tot}} - C) \right),$$

$$\frac{dC}{dt} = -\bar{G} \left(\gamma C - \frac{\gamma}{K_d} X^* (p_{\text{tot}} - C) \right).$$

This system is in the form of system (6.23) with $u = Z$, $x = (C_1, Z^*, X^*)$, and $v = C$, so that we can choose T as the 3×3 identity matrix and

$$M = \begin{pmatrix} 0 \\ 0 \\ 1 \end{pmatrix}.$$

The reader can verify that the Jacobian of $f_1(x, u)$ has all eigenvalues with negative real part (Exercise 6.10).

Figure 6.18a shows that, for a periodic input $k(t)$, the system with low value for G suffers from retroactivity to the output. However, for a large value of G (Figure 6.18b), the permanent behavior of the connected system becomes similar to that of the isolated system, whether $G \gg \bar{G}$, $G = \bar{G}$ or $G \ll \bar{G}$. This confirms the theoretical result that, independently of the order of magnitude of $\bar{G}$, the system can arbitrarily attenuate retroactivity for large enough G. Note that this robustness to the load applied on X^* is achieved even if the concentration of X^* is about 100 times smaller than the concentration of the load applied to it. This allows us to design the system such that it can output any desired value while being robust to retroactivity.

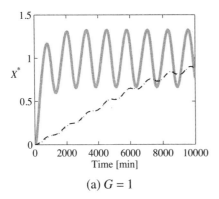

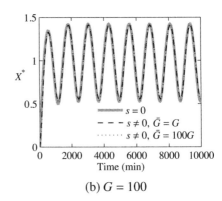

(a) $G = 1$ (b) $G = 100$

Figure 6.18: Output response of the phosphotransfer system with a periodic signal $k(t) = \gamma(1 + \sin \omega t)$. The parameters are given by $\gamma = 0.01$ min^{-1}, $X_{\text{tot}} = 5000$ nM, $k_1 = k_2 = k_3 = k_4 = 0.01$ min^{-1} nM^{-1}, $\pi_1 = \pi_2 = 0.01G$ min^{-1} in which $G = 1$ in (a), and $G = 100$ in (b). The downstream system parameters are given by $K_{\text{d}} = 1$ nM and $k_{\text{off}} = 0.01\bar{G}$ min^{-1}, in which $\bar{G}$ takes the values indicated in the legend. The isolated system ($s = 0$) corresponds to $p_{\text{tot}} = 0$ while the connected system ($s \neq 0$) corresponds to $p_{\text{tot}} = 100$ nM.

6.6   A case study on the use of insulation devices

In this section, we consider again the problem illustrated at the beginning of the chapter in which we would like to transmit the periodic stimulation of the activator-repressor clock to a downstream system (Figure 6.2b). We showed before that connecting the clock directly to the downstream system causes the oscillations to be attenuated and even quenched (Figure 6.2c), so that we fail to transmit the desired periodic stimulation to the downstream system. Here, we describe a solution to this problem that implements an insulation device to connect the clock to the downstream system. This way, the downstream system receives the desired periodic input stimulation despite the potentially large retroactivity s that this system applies to the insulation device. In particular, we employ the insulation device realized by a phosphorylation cycle in the configuration shown in Figure 6.19. The top diagram illustrates a simplified genetic layout of the clock. The activator A is expressed from a gene under the control of a promoter activated by A and repressed by B, while the repressor is expressed from a gene under the control of a promoter activated by A. Protein A, in turn, activates the expression of protein D in the downstream system. In this case, the promoter p controlling the expression of D contains operator regions that A recognizes, so that A can bind to it.

When the insulation device of Figure 6.14 is employed to interconnect the clock to the downstream system, two modifications need to be made to enable the connections. Since A is not a kinase, we need to insert downstream of the gene expressing A another gene expressing the kinase Z (bottom diagram of Figure 6.19). Since both A and Z are under the control of the same promoter, they will be pro-

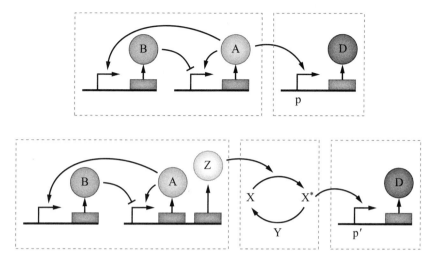

Figure 6.19: The activator-repressor clock connected to its downstream system through the insulation device of Figure 6.14. The top diagram illustrates a simplified genetic layout of the activator-repressor clock of Figure 6.2a. The bottom diagram illustrates how the genetic layout of the clock should be modified such that it can connect to the phosphorylation cycle that takes as input the kinase Z. In this case, the downstream system still expresses protein D, but its expression is controlled by a different promoter that is activated by X^* as opposed to being activated by A.

duced at the same rates and hence the concentration of Z should mirror that of A if the decay rates are the same for both proteins. Note that a solution in which we insert downstream of the gene expressing A a transcription factor Z that directly binds to downstream promoter sites p' to produce D (without the insulation device in between) would not solve the problem. In fact, while the clock behavior would be preserved in this case, the behavior of the concentration of Z would not mirror

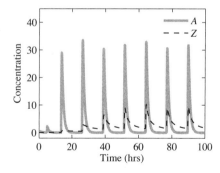

Figure 6.20: Simulation results for the concentration of protein Z in Figure 6.19 in the case in which this were used directly as an input to the downstream system, thus binding sites p' (dashed plot). The clock parameters are the same as those in Figure 6.2c and $\gamma_K = \gamma_A$.

that of A since protein Z would be loaded by the downstream promoter sites p′ (Figure 6.20). As a consequence, we would still fail to transmit the clock signal $A(t)$ to protein D. The second modification that needs to be made is to change the promoter p to a new promoter p′ that has an operator that protein X^* recognizes (bottom diagram of Figure 6.19).

In the case of the bottom diagram of Figure 6.19, the dynamics of the clock proteins remain the same as that of model (5.11) and given by

$$\frac{dA}{dt} = \frac{\kappa_A}{\delta_A} \frac{\alpha_A(A/K_A)^n + \alpha_{A0}}{1 + (A/K_A)^n + (B/K_B)^m} - \gamma_A A,$$

$$\frac{dB}{dt} = \frac{\kappa_B}{\delta_B} \frac{\alpha_B(A/K_A)^n + \alpha_{B0}}{1 + (A/K_A)^n} - \gamma_B B.$$

To these equations, we need to add the dynamics of the kinase $Z(t)$, which, when the phosphorylation cycle is not present, will be given by

$$\frac{dZ}{dt} = \frac{\kappa_A}{\delta_A} \frac{\alpha_A(A/K_A)^n + \alpha_{A0}}{1 + (A/K_A)^n + (B/K_B)^m} - \gamma_Z Z. \tag{6.37}$$

Note that we are using for simplicity the two-dimensional model of the activator-repressor clock. A similar result would be obtained using the four-dimensional model that incorporates the mRNA dynamics (see Exercise 6.13).

When the phosphorylation cycle is present, the differential equation for Z given by (6.37) changes to

$$\frac{dZ}{dt} = \frac{\kappa_A}{\delta_A} \frac{\alpha_A(A/K_A)^n + \alpha_{A0}}{1 + (A/K_A)^n + (B/K_B)^m} - \gamma_Z Z$$
$$- \left[a_1 X_{\text{tot}} Z \left(1 - \frac{X^*}{X_{\text{tot}}} - \frac{C_1}{X_{\text{tot}}} - \frac{C_2}{X_{\text{tot}}} - \frac{C}{X_{\text{tot}}} \right) - (d_1 + k_1)C_1 \right], \tag{6.38}$$

in which the term in the square brackets is the retroactivity to the input r of the insulation device. The model of the insulation device with the downstream system remains the same as before and given by equations (6.26)–(6.29).

Figure 6.21 shows the trajectories of $Z(t)$, and $X^*(t)$ for the system of Figure 6.19. As desired, the signal $X^*(t)$, which drives the downstream system, closely tracks $A(t)$ plotted in Figure 6.20 despite the retroactivity due to load applied by the downstream sites p′. Note that the trajectory of $Z(t)$ is essentially the same whether the insulation device is present or not, indicating a low retroactivity to the input r. The retroactivity to the output s only slightly affects the output of the insulation device (Figure 6.21b). The plot of Figure 6.21b, showing the signal that drives the downstream system, can be directly compared to the signal that would drive the downstream system in the case in which the insulation device would not be used (Figure 6.20, dashed plot). In the latter case, the downstream system would not be properly driven, while with the insulation device it is driven as expected even in the face of a large load.

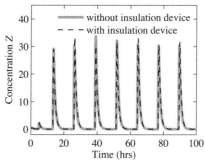
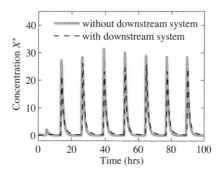

(a) Input of the insulation device   (b) Output of the insulation device

Figure 6.21: Simulation results for the system of Figure 6.19. Panel (a) shows the concentration of the kinase Z without ($r = 0$ in equation (6.25)) and with the insulation device. Panel (b) shows the behavior of the output of the insulation device X^* without ($s = 0$) and with the downstream system. The clock parameters are the same as those in Figure 6.2c and $\gamma_K = \gamma_A$. The phosphorylation cycle parameters are as in Figure 6.16a. The load parameters are given by $k_{\mathrm{on}} = 50$ min^{-1} nM^{-1}, $k_{\mathrm{off}} = 50$ min^{-1}, and $p_{\mathrm{tot}} = 100$ nM.

Exercises

6.1 Include in the study of retroactivity in transcriptional systems the mRNA dynamics and demonstrate how/whether the results change. Specifically, consider the following model of a connected transcriptional system:

$$\frac{dm_X}{dt} = k(t) - \delta m_X,$$

$$\frac{dX}{dt} = \kappa m_X - \gamma X + [k_{\mathrm{off}}C - k_{\mathrm{on}}(p_{\mathrm{tot}} - C)X],$$

$$\frac{dC}{dt} = -k_{\mathrm{off}}C + k_{\mathrm{on}}(p_{\mathrm{tot}} - C)X.$$

6.2 Consider the connected transcriptional system model in standard singular perturbation form with $\epsilon \ll 1$:

$$\frac{dz}{dt} = k(t) - \gamma(z - C), \qquad \epsilon\frac{dC}{dt} = -\gamma C + \frac{\gamma}{k_d}(p_{\mathrm{tot}} - C)(z - C).$$

Demonstrate that the slow manifold is locally asymptotically stable.

6.3 The characterization of retroactivity effects in a transcriptional module was based on the following model of the interconnection:

$$\frac{dX}{dt} = k(t) - \gamma X + [k_{\mathrm{off}}C - k_{\mathrm{on}}(p_{\mathrm{tot}} - C)X],$$

$$\frac{dC}{dt} = -k_{\mathrm{off}}C + k_{\mathrm{on}}(p_{\mathrm{tot}} - C)X,$$

in which the dilution of the complex C was neglected. This is often a fair assumption, but depending on the experimental conditions, a more appropriate model may include dilution for the complex C. In this case, the model modifies to

$$\frac{dX}{dt} = k(t) - (\mu + \bar{\gamma})X + [k_{\text{off}}C - k_{\text{on}}(p_{\text{tot}} - C)X],$$

$$\frac{dC}{dt} = -k_{\text{off}}C + k_{\text{on}}(p_{\text{tot}} - C)X - \mu C,$$

in which μ represents decay due to dilution and $\bar{\gamma}$ represents protein degradation. Employ singular perturbation to determine the reduced X dynamics and the effects of retroactivity in this case. Is the steady state characteristic curve of the transcription module affected by retroactivity? Determine the extent of this effect as $\mu/\bar{\gamma}$ decreases.

6.4 In this problem, we study the frequency dependent effects of retroactivity in gene circuits through simulation to validate the findings obtained through linearization in Section 6.3. In particular, consider the model of a connected transcriptional component (6.3). Consider the parameters provided in Figure 6.5 and simulate the system with input $k(t) = \gamma(1 + sin(\omega t))$ with $\omega = 0.005$. Then, decrease and increase the frequency progressively and make a frequency/amplitude plot for both connected and isolated systems. Increase γ and redo the frequency/amplitude plot. Comment on the retroactivity effects that you observe.

6.5 Consider the negatively autoregulated gene illustrated in Section 5.2. Instead of modeling negative autoregulation using the Hill function, explicitly model the binding of A with its own promoter. In this case, the formed complex C will be transcriptionally inactive (see Section 2.3). Explore through simulation how the response of the system without negative regulation compares to that with negative regulation when the copy number of the A gene is increased and the unrepressed expression rate β is decreased.

6.6 We have illustrated that the expression of the point of half-maximal induction in a covalent modification cycle is affected by the effective load λ as follows:

$$y_{50} = \frac{\bar{K}_1 + 0.5}{\bar{K}_2(1 + \lambda) + 0.5}.$$

Study the behavior of this quantity when the effective load λ is changed.

6.7 Show how equation (6.9) is derived in Section 6.4.

6.8 Demonstrate through a mathematical proof that in the following system

$$\frac{dX}{dt} = G(k(t) - KX)(1 - d(t)),$$

in which $0 < d(t) < 1$ and $|\dot{k}(t)|$ is bounded, we have that $X(t) - k(t)/K$ becomes smaller as G is increased.

6.9 Consider the one-step reaction model of the phosphorylation cycle with downstream binding sites given in equation (6.20). Simulate the system and determine how the behavior of the connected system compares to that of the isolated system when the amounts of substrate and phosphatase X_{tot} and Y_{tot} are increased.

6.10 Demonstrate that the Jacobian $\partial f_1(x, u)/\partial x$ for the system in equations (6.30) has eigenvalues with negative real part. You can demonstrate this by using symbolic computation, or you can use the parameter values of Figure 6.16.

6.11 Demonstrate that the Jacobian $\partial f_1(x, u)/\partial x$ for the system in equations (6.36) has eigenvalues with negative real part. You can demonstrate this by using symbolic computation, or you can use the parameter values of Figure 6.18.

6.12 Consider the activator-repressor clock described in Section 5.5 and take the parameter values of Figure 5.9 that result in a limit cycle. Then, assume that the activator A connects to another transcriptional circuit through the cooperative binding of n copies of A with operator sites p to form the complex C:

$$nA + p \underset{k_{off}}{\overset{k_{on}}{\rightleftharpoons}} C$$

with conservation law $p + C = p_{tot}$ (connected clock). Answer the following questions:

  (i) Simulate the connected clock and vary the total amount of p, that is, p_{tot}. Explore how this affects the behavior of the clock.

 (ii) Give a mathematical explanation of the phenomenon you saw in (i). To do so, use singular perturbation to approximate the dynamics of the clock with downstream binding on the slow manifold (here, $k_{off} \gg \gamma_A, \gamma_B$).

(iii) Assume now that A does not bind to sites p, while the repressor B does. Take the parameter values of Figure 5.9 that result in a stable equilibrium. Explore how increasing p_{tot} affects the clock trajectories.

6.13 Consider the system depicted in Figure 6.19 and model the activator-repressor clock including the mRNA dynamics as shown in Section 5.5. Demonstrate through simulation that the same results obtained in Section 6.6 with a two-dimensional model of the clock still hold.

Chapter 7
Design Tradeoffs

In this chapter we describe some of the design tradeoffs arising from the interaction between synthetic circuits and the host organism. We specifically focus on two issues. The first issue is concerned with the effects of competition for shared cellular resources on circuits' behavior. In particular, circuits (endogenous and exogenous) share a number of cellular resources, such as RNA polymerase, ribosomes, ATP, enzymes, and nucleotides. The insertion or induction of synthetic circuits in the cellular environment changes the for these resources, with possibly undesired repercussions on the functioning of the circuits. Independent circuits may become coupled when they share common resources that are not in overabundance. This fact leads to constraints among the concentrations of proteins in synthetic circuits, which should be accounted for in the design phase. The second issue we consider is the effect of biological noise on the design of devices requiring high gains. Specifically, we illustrate possible design tradeoffs between retroactivity attenuation and noise amplification that emerge due to the intrinsic noise of biomolecular reactions.

7.1 Competition for shared cellular resources

Exogenous circuits, just like endogenous ones, use cellular resources—such as ribosomes, RNA polymerase (RNAP), enzymes and ATP—that are shared among all the circuitry of the cell. From a signals and systems point of view, these interactions can be depicted as in Figure 7.1. The cell's endogenous circuitry produces resources as output and exogenous synthetic circuits take these resources as inputs. As a consequence, as seen in Chapter 6, there is retroactivity from the exogenous circuits to the cellular resources. This retroactivity creates indirect coupling between the exogenous circuits and can lead to undesired crosstalk. In this chapter, we study the effect of the retroactivity from the synthetic circuits to shared resources in the cellular environment by focusing on the effect on availability of RNA polymerase and ribosomes, for simplicity. We then study the consequence of this retroactivity, illustrating how the behavior of individual circuits becomes coupled. These effects are significant for any resource whose availability is not in substantial excess compared to the demand by exogenous circuits.

In order to illustrate the problem, we consider the simple system shown in Figure 7.2, in which two modules, a constitutively expressed gene (Module 1) and a gene activated by a transcriptional activator A (Module 2), are present in the cel-

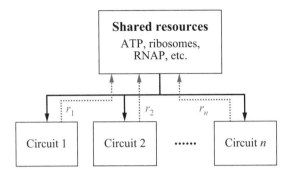

Figure 7.1: The cellular environment provides resources to synthetic circuits, such as RNA polymerase, ribosomes, ATP, nucleotides, proteases, etc. These resources can be viewed as an "output" of the cell's endogenous circuitry and an input to the exogenous circuits. Circuit i takes these resources as input and, as a consequence, it causes a retroactivity r_i to its input. Hence, the endogenous circuitry has a retroactivity to the output s that encompasses all the retroactivities applied by the exogenous circuits.

lular environment. In theory, Module 2 should respond to changes in the activator A concentration, while Module 1, having a constitutively active promoter, should display a constant expression level that is independent of the activator A concentration. Experimental results, however, indicate that this is not the case: Module 1's output protein concentration P_1 also responds to changes in the activator A concentration. In particular, as the activator A concentration is increased, the concentration of protein P_1 can substantially decrease. This fact can be qualitatively explained by the following reasoning. When A is added, RNA polymerase can bind to DNA promoter D_2 and start transcription, so that the free available RNA polymerase decreases as some is bound to the promoter D_2. Transcription of Module 2 generates mRNA and hence ribosomes will have more ribosome binding sites to which they can bind, so that less ribosomes will be free and available for other reactions. It follows that the addition of activator A leads to an overall decrease of the free RNA polymerase and ribosomes that can take part in the transcription and translation reactions of Module 1. The net effect is that less of P_1 protein will be produced.

The extent of this effect will depend on the overall availability of the shared resources, on the biochemical parameters, and on whether the resources are regulated. For example, it is known that ribosomes are internally regulated by a combination of feedback interactions [61]. This, of course, may help compensate for changes in the demand of these resources, though experiments demonstrate that the coupling effects are indeed noticeable [98].

In this chapter, we illustrate how this effect can be mathematically explained by explicitly accounting for the usage of RNA polymerase and ribosomes in the transcription and translation models of the circuits. To simplify the mathematical analysis and to gather analytical understanding of the key parameters at the basis of

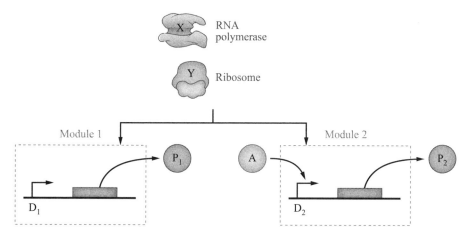

Figure 7.2: Module 1 has a constitutively active promoter that controls the expression of protein P_1, while Module 2 has a promoter activated by activator A, which controls the expression of protein P_2. The two modules do not share any transcription factors, so they are not "connected." Both of them use RNA polymerase (X) and ribosomes (Y) for the transcription and translation processes.

this phenomenon, we first focus on the usage of RNA polymerase, neglecting the usage of ribosomes. We then provide a computational model that accounts for both RNA polymerase and ribosome utilization and illustrate quantitative simulation results.

Analytical study

To illustrate the essence of the problem, we assume that gene expression is a one-step process, in which the RNA polymerase binds to the promoter region of a gene resulting in a transcriptionally active complex that, in turn, produces the protein. That is, we will be using the lumped reactions (2.12), in which on the right-hand side of the reaction we have the protein instead of the mRNA.

By virtue of this simplification, we can write the reactions describing Module 1 as

$$X+D_1 \underset{d_1}{\overset{a_1}{\rightleftharpoons}} X{:}D_1 \xrightarrow{k_1} P_1 + X + D_1, \qquad P_1 \xrightarrow{\gamma} \emptyset.$$

The reactions describing Module 2 can be written similarly, recalling that in the presence of an activator they should be modified according to equation (2.21). Taking this into account, the reactions of Module 2 are given by

$$D_2+A \underset{d_0}{\overset{a_0}{\rightleftharpoons}} D_2{:}A, \qquad X+D_2{:}A \underset{d_2}{\overset{a_2}{\rightleftharpoons}} X{:}D_2{:}A \xrightarrow{k_2} P_2 + X + D_2{:}A, \qquad P_2 \xrightarrow{\gamma} \emptyset.$$

We let $D_{\text{tot},1}$ and $D_{\text{tot},2}$ denote the total concentration of DNA for Module 1 and Module 2, respectively, and we let $K_0 = d_0/a_0$, $K_1 = d_1/a_1$, and $K_2 = d_2/a_2$. By approximating the complexes' concentrations by their quasi-steady state values, we obtain the expressions

$$[X{:}D_1] = D_{\text{tot},1} \frac{X/K_1}{1 + X/K_1}, \qquad [X{:}D_2{:}A] = D_{\text{tot},2} \frac{(A/K_0)(X/K_2)}{1 + (A/K_0)(1 + X/K_2)}. \quad (7.1)$$

As a consequence, the differential equation model for the system is given by

$$\frac{dP_1}{dt} = k_1 D_{\text{tot},1} \frac{X/K_1}{1 + X/K_1} - \gamma P_1,$$
$$\frac{dP_2}{dt} = k_2 D_{\text{tot},2} \frac{(A/K_0)(X/K_2)}{1 + (A/K_0)(1 + X/K_2)} - \gamma P_2,$$

so that the steady state values of P_1 and P_2 are given by

$$P_1 = \frac{k_1 D_{\text{tot},1}}{\gamma} \frac{X/K_1}{1 + X/K_1}, \qquad P_2 = \frac{k_2 D_{\text{tot},2}}{\gamma} \frac{(A/K_0)(X/K_2)}{1 + (A/K_0)(1 + X/K_2)}.$$

These values are indirectly coupled through the conservation law of RNA polymerase. Specifically, we let X_{tot} denote the total concentration of RNA polymerase. This value is mainly determined by the cell growth rate and for a given growth rate it is about constant [15]. Then, we have that $X_{\text{tot}} = X + [X{:}D_1] + [X{:}D_2{:}A]$, which, considering the expressions of the quasi-steady state values of the complexes' concentrations in equation (7.1), leads to

$$X_{\text{tot}} = X + D_{\text{tot},1} \frac{X/K_1}{1 + X/K_1} + D_{\text{tot},2} \frac{(A/K_0)(X/K_2)}{1 + (A/K_0)(1 + X/K_2)}. \quad (7.2)$$

We next study how the steady state value of X is affected by the activator concentration A and how this effect is reflected in a dependency of P_1 on A. To perform this study, it is useful to write $\alpha := (A/K_0)$ and note that for α sufficiently small (sufficiently small amounts of activator A), we have that

$$\frac{\alpha(X/K_2)}{1 + \alpha(1 + X/K_2)} \approx \alpha(X/K_2).$$

Also, to simplify the derivations, we assume that the binding of X to D_1 is sufficiently weak, that is, $X \ll K_1$. In light of this, we can rewrite the conservation law (7.2) as

$$X_{\text{tot}} = X + D_{\text{tot},1} \frac{X}{K_1} + D_{\text{tot},2} \alpha \frac{X}{K_2}.$$

This equation can be explicitly solved for X to yield

$$X = \frac{X_{\text{tot}}}{1 + (D_{\text{tot},1}/K_1) + \alpha(D_{\text{tot},2}/K_2)}.$$

This expression depends on α, and hence on the activator concentration A. Specifically, as the activator is increased, the value of the free X concentration monotonically decreases. As a consequence, the equilibrium value P_1 will also depend on A according to

$$P_1 = \frac{k_1 D_{\text{tot},1}}{\gamma} \frac{X_{\text{tot}}/K_1}{1 + (D_{\text{tot},1}/K_1) + \alpha(D_{\text{tot},2}/K_2)},$$

so that P_1 monotonically decreases as A is increased. That is, Module 1 responds to changes in the activator of Module 2. From these expressions, we can also deduce that if $D_{\text{tot},1}/K_1 \gg \alpha D_{\text{tot},2}/K_2$, that is, the demand for RNA polymerase in Module 1 is much larger than that of Module 2, then changes in the activator concentration will lead to small changes in the free amount of RNA polymerase and in P_1.

This analysis illustrates that forcing an increase in the expression of any protein causes an overall decrease in available resources, which leads to a decrease in the expression of other proteins. As a consequence, there is a tradeoff between the amount of protein produced by one circuit and the amount of proteins produced by different circuits. In addition to a design tradeoff, this analysis illustrates that "unconnected" circuits can affect each other because they share common resources. This can, in principle, lead to a dramatic departure of a circuit's behavior from its nominal one. As an exercise, the reader can verify that similar results hold in the case in which Module 2 has a repressible promoter instead of one that can be activated (see Exercise 7.2).

The model that we have presented here contains many simplifications. In addition to the mathematical approximations performed and to the fact that we did not account for ribosomes, the model neglects the transcription of endogenous genes. In fact, RNA polymerase is also used for transcription of chromosomal genes. While the qualitative behavior of the coupling between Module 1 and Module 2 is not going to be affected by including endogenous transcription, the extent of this coupling may be substantially impacted. In the next section, we illustrate how the presence of endogenous genes may affect the extent to which the availability of RNA polymerase decreases upon addition of exogenous genes.

Estimates of RNA polymerase perturbations by exogenous plasmids

In the previous section, we illustrated the mechanism by which the change in the availability of a shared resource, due to the addition of synthetic circuits, can cause crosstalk between unconnected circuits. The extent of this crosstalk depends on the amount by which the shared resource changes. This amount, in turn, depends on the specific values of the dissociation constants, the total resource amounts, and the fraction of resource that is used already by natural circuits. Here, we consider how the addition of an external plasmid affects the availability of RNA polymerase, considering a simplified model of the interaction of RNA polymerase with the exogenous and natural DNA.

In *E. coli*, the amount of RNA polymerase and its partitioning mainly depends on the growth rate of the cell [15]: with 0.6 doublings/hour there are only 1500 molecules/cell, while with 2.5 doublings/hour this number is 11400. The fraction of active RNA polymerase molecules also increases with the growth rate. For illustration purposes, we assume here that the growth rate is the highest considered in [15], so that 1 molecule/cell corresponds to approximately 1nM concentration. In this case, a reasonable partitioning of the total pool of RNA polymerase of concentration $X_{tot} = 12\ \mu M$ is the following [57]:

(i) specifically DNA-bound (at promoter) X_s: 30% (4000 molecules/cell, that is, $X_s = 4\ \mu M$),

(ii) non-specifically DNA-bound X_n: 60% (7000 molecules/cell, that is, $X_n = 7\ \mu M$),

(iii) free X: 10% (1000 molecules/cell, that is, $X = 1\ \mu M$).

By [16], the number of initiations per promoter can be as high as 30/minute in the case of constitutive promoters, and 1-3/minute for regulated promoters. Here, we choose an effective value of 5 initiations/minute per promoter, so that on average, 5 molecules of RNA polymerase can be simultaneously transcribing each gene, as transcribing a gene takes approximately a minute [4]. There are about 1000 genes expressed in exponential growth phase [47], hence we approximate the number of promoter binding sites for X to 5000, or $D_{tot} = 5\ \mu M$. The binding reaction for specific binding is of the form

$$D + X \overset{a}{\underset{d}{\rightleftharpoons}} D{:}X,$$

in which D represents DNA promoter binding sites DNA^p in total concentration D_{tot}. Consequently, we have $D_{tot} = D + [D{:}X]$. At the equilibrium, we have $[D{:}X] = X_s = 4\ \mu M$ and $D = D_{tot} - [D{:}X] = D_{tot} - X_s = 1\ \mu M$. With dissociation constant $K_d = d/a$ the equilibrium is given by $0 = DX - K_d[D{:}X]$, hence we have that $K_d = DX/[D{:}X] = 0.25\ \mu M$, which can be interpreted as an "effective" dissociation constant. This is in the range $1\ nM - 1\ \mu M$ suggested by [38] for specific binding of RNA polymerase to DNA. Therefore, we are going to model the specific binding of RNA polymerase to the chromosome of *E. coli* in exponential growth phase as one site with concentration D_{tot} and effective dissociation constant K_d.

Furthermore, we have to take into account the rather significant amount of RNA polymerase bound to the DNA other than at the promoter region ($X_n = 7\ \mu M$). To do so, we follow a similar path as in the case of specific binding. In particular, we model the non-specific binding of RNA polymerase to DNA as

$$\bar{D} + X \overset{\bar{a}}{\underset{\bar{d}}{\rightleftharpoons}} \bar{D}{:}X,$$

in which $\bar{D}$ represents DNA binding sites with concentration $\bar{D}_{tot}$ and effective dissociation constant $\bar{K}_d = \bar{d}/\bar{a}$. At the equilibrium, we have that the concentration of RNA polymerase non-specifically bound to DNA is given by

$$X_n = [\bar{D}:X] = \frac{\bar{D}_{tot}X}{X + \bar{K}_d}.$$

As the dissociation constant $\bar{K}_d$ of non-specific binding of RNA polymerase to DNA is in the range $1 - 1000 \ \mu M$ [38], we have $X \ll \bar{K}_d$, yielding $X_n = [\bar{D}:X] \approx X\bar{D}_{tot}/\bar{K}_d$. Consequently, we obtain $\bar{D}_{tot}/\bar{K}_d = X_n/X = 7$. Here, we did not model the reaction in which the non-specifically bound RNA polymerase X_n slides to the promoter binding sites D. This would not substantially affect the results of our calculations because the RNA polymerase non-specifically bound on the chromosome cannot bind the plasmid promoter sites without first unbinding and becoming free.

Now, we can consider the addition of synthetic plasmids. Specifically, we consider high-copy number plasmids (copy number $100 - 300$) with one copy of a gene under the control of a constitutive promoter. We abstract it by a binding site for RNA polymerase D′ to which X can bind according to the following reaction:

$$D' + X \underset{d'}{\overset{a'}{\rightleftharpoons}} D':X,$$

where D′ is the RNA polymerase-free binding site and D′ : X is the site bound to RNA polymerase. Consequently, we have $D'_{tot} = D' + [D':X]$, where $D'_{tot} = 1 \ \mu M$, considering 200 copies of plasmid per cell and 5 RNA polymerase molecules per gene. The dissociation constant corresponding to the above reaction is given by $K' = d'/a'$. At the steady state we have

$$[D':X] = D'_{tot}\frac{X}{K'_d + X},$$

together with the conservation law for RNA polymerase given by

$$X + [D:X] + [\bar{D}:X] + [D':X] = X_{tot}. \tag{7.3}$$

In this model, we did not account for RNA polymerase molecules paused or queuing on the chromosome; moreover, we also neglected the resistance genes on the plasmid and all additional sites (specific or not) to which RNA polymerase can also bind. Hence, we are underestimating the effect of load presented by the plasmid.

Solving equation (7.3) for the free RNA polymerase amount X gives the following results. These results depend on the ratio between the effective dissociation constant K_d of RNA polymerase binding with the natural DNA promoters and the dissociation constant K'_d of binding with the plasmid promoter:

(i) $K'_d = 0.1 K_d$ (RNA polymerase binds stronger to the plasmid promoter) results in $X = 0.89 \ \mu M$, that is, the concentration of free RNA polymerase decreases by about 11%;

(ii) $K'_d = K_d$ (binding is the same) results in $X = 0.91$ μM, consequently, the concentration of free RNA polymerase decreases by about 9%;

(iii) $K'_d = 10K_d$ (RNA polymerase binds stronger to the chromosome) results in $X = 0.97$ μM, which means that the concentration of free RNA polymerase decreases by about 3%.

Note that the decrease in the concentration of free RNA polymerase is greatly reduced by the significant amount of RNA polymerase being non-specifically bound to the DNA. For instance, in the second case when $K'_d = K_d$, the RNA polymerase molecules sequestered by the synthetic plasmid can be partitioned as follows: about 10% is taken from the pool of free RNA polymerase molecules X, another 10% comes from the RNA polymerase molecules specifically bound, and the overwhelming majority (80%) decreases the concentration of RNA polymerase non-specifically bound to DNA. That is, this weak binding of RNA polymerase to DNA acts as a buffer against changes in the concentration of free RNA polymerase.

We conclude that if the promoter on the synthetic plasmid has a dissociation constant for RNA polymerase that is in the range of the dissociation constant of specific binding, the perturbation on the available free RNA polymerase is about 9%. This perturbation, even if fairly small, may in practice result in large effects on the protein concentration. This is because it may cause a large perturbation in the concentration of free ribosomes. In fact, one added copy of an exogenous plasmid will lead to transcription of several mRNA molecules, which will demand ribosomes for translation. Hence, a small increase in the demand for RNA polymerase may be associated with a dramatically larger increase in the demand for ribosomes. This is illustrated in the next section through a computational model including ribosome sharing.

Computational model and numerical study

In this section, we introduce a model of the system in Figure 7.2, in which we consider both the RNA polymerase and the ribosome usage. We let the concentration of RNA polymerase be denoted by X and the concentration of ribosomes be denoted by Y. We let m_1 and P_1 denote the concentrations of the mRNA and protein in Module 1 and let m_2 and P_2 denote the concentrations of the mRNA and protein in Module 2. The reactions of the transcription process in Module 1 are given by (see Section 2.2)

$$X + D_1 \underset{d_1}{\overset{a_1}{\rightleftharpoons}} X{:}D_1 \xrightarrow{k_1} m_1 + X + D_1, \quad m_1 \xrightarrow{\delta} \emptyset,$$

while the translation reactions are given by

$$Y + m_1 \underset{d'_1}{\overset{a'_1}{\rightleftharpoons}} Y{:}m_1 \xrightarrow{k'_1} P_1 + m_1 + Y, \quad Y{:}m_1 \xrightarrow{\delta} Y, \quad P_1 \xrightarrow{\gamma} \emptyset.$$

The resulting system of differential equations is given by

$$\frac{d}{dt} [\text{X:D}_1] = a_1 X D_1 - (d_1 + k_1) [\text{X:D}_1],$$

$$\frac{dm_1}{dt} = k_1 [\text{X:D}_1] - a_1' Y m_1 + d_1' [\text{Y:m}_1] - \delta m_1 + k_1' [\text{Y:m}_1], \qquad (7.4)$$

$$\frac{d}{dt} [\text{Y:m}_1] = a_1' Y m_1 - (d_1' + k_1;) [\text{Y:m}_1],$$

$$\frac{dP_1}{dt} = k_1' [\text{Y:m}_1] - \gamma P_1,$$

in which $D_1 = D_{\text{tot},1} - [\text{X:D}_1]$ from the conservation law of DNA in Module 1.

The reactions of the transcription process in Module 2 are given by (see Section 2.3)

$$\text{D}_2 + \text{A} \underset{d_0}{\overset{a_0}{\rightleftharpoons}} \text{D}_2\text{:A}, \quad \text{X} + \text{D}_2\text{:A} \underset{d_2}{\overset{a_2}{\rightleftharpoons}} \text{X:D}_2\text{:A} \overset{k_2}{\rightarrow} m_2 + \text{X} + \text{D}_2\text{:A}, \quad m_2 \overset{\delta}{\rightarrow} \emptyset,$$

while the translation reactions are given by

$$\text{Y} + m_2 \underset{d_2'}{\overset{a_2'}{\rightleftharpoons}} \text{Y:m}_2 \overset{k_2'}{\rightarrow} \text{P}_2 + m_2 + \text{Y}, \quad \text{Y:m}_2 \overset{\delta}{\rightarrow} \text{Y}, \quad \text{P}_2 \overset{\gamma}{\rightarrow} \emptyset.$$

The resulting system of differential equations is given by

$$\frac{d}{dt} [\text{D}_2\text{:A}] = a_0 D_2 A - d_0 [\text{D}_2\text{:A}] - a_2 X [\text{D}_2\text{:A}] + (d_2 + k_2)[\text{X:D}_2\text{:A}],$$

$$\frac{d}{dt} [\text{X:D}_2\text{:A}] = a_2 X [\text{D}_2\text{:A}] - (d_2 + k_2) [\text{X:D}_2\text{:A}],$$

$$\frac{dm_2}{dt} = k_2 [\text{X:D}_2\text{:A}] - a_2' Y m_2 + d_2' [\text{Y:m}_2] - \delta m_2 + k_2' [\text{Y:m}_2], \qquad (7.5)$$

$$\frac{d}{dt} [\text{Y:m}_2] = a_2' Y m_2 - (d_2' + k_2') [\text{Y:m}_2] - \delta [\text{Y:m}_2],$$

$$\frac{dP_2}{dt} = k_2' [\text{Y:m}_2] - \gamma P_2,$$

in which we have that $D_2 = D_{\text{tot},2} - [\text{D}_2\text{:A}] - [\text{X:D}_2\text{:A}]$ by the conservation law of DNA in Module 2.

The two modules are coupled by the conservation laws for RNA polymerase and ribosomes given by

$$X_{\text{tot}} = X + [\text{X:D}_1] + [\text{X:D}_2\text{:A}], \quad Y_{\text{tot}} = Y + [\text{Y:m}_1] + [\text{Y:m}_2],$$

which we employ in systems (7.4)–(7.5) by writing

$$X = X_{\text{tot}} - [\text{X:D}_1] - [\text{X:D}_2\text{:A}], \quad Y = Y_{\text{tot}} - [\text{Y:m}_1] - [\text{Y:m}_2].$$

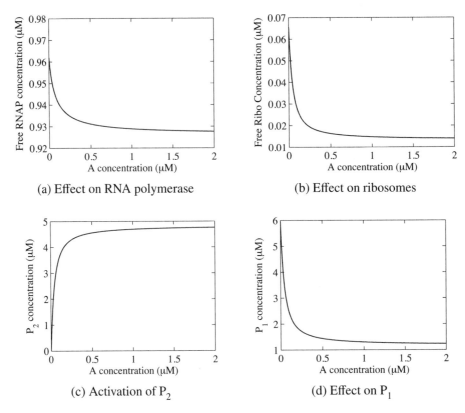

(a) Effect on RNA polymerase

(b) Effect on ribosomes

(c) Activation of P_2

(d) Effect on P_1

Figure 7.3: Simulation results for the ordinary differential equation model (7.4)–(7.5). When A is increased, X slightly decreases (a) while Y decreases substantially (b). So, as P_2 increases (c), we have that P_1 decreases substantially (d). For this model, the parameter values were taken from http://bionumbers.hms.harvard.edu as follows. For the concentrations, we have set $X_{\text{tot}} = 1\ \mu\text{M}$, $Y_{\text{tot}} = 10\ \mu\text{M}$, and $D_{\text{tot},1} = D_{\text{tot},2} = 0.2\ \mu\text{M}$. The values of the association and dissociation rate constants were chosen such that the corresponding dissociation constants were in the range of dissociation constants for specific binding. Specifically, we have $a_0 = 10\ \mu\text{M}^{-1}\text{min}^{-1}$, $d_0 = 1\ \text{min}^{-1}$, $a_2 = 10\ \mu\text{M}^{-1}\text{min}^{-1}$, $d_2 = 1\ \text{min}^{-1}$, $a'_2 = 100\ \mu\text{M}^{-1}\text{min}^{-1}$, $d'_2 = 1\ \text{min}^{-1}$, $a_1 = 10\ \mu\text{M}^{-1}\text{min}^{-1}$, $d_1 = 1\ \text{min}^{-1}$, $a'_1 = 10\ \mu\text{M}^{-1}\text{min}^{-1}$, and $d'_1 = 1\ \text{min}^{-1}$. The transcription and translation rate constants were chosen to give a few thousands of protein copies per cell and calculated considering the elongation speeds, the average length of a gene, and the average number of RNA polymerase per gene and of ribosomes per transcript. The resulting values chosen are given by $k_1 = k_2 = 40\ \text{min}^{-1}$ and $k'_1 = k'_2 = 0.006\ \text{min}^{-1}$. Finally, the decay rates are given by $\gamma = 0.01\ \text{min}^{-1}$ corresponding to a protein half life of about 70 minutes and $\delta = 0.1\ \text{min}^{-1}$ corresponding to an mRNA half life of about 7 minutes.

Simulation results are shown in Figure 7.3a–7.3d, in which we consider cells growing at high rate. In the simulations, we have chosen $X_{\text{tot}} = 1\ \mu\text{M}$ to account for the fact that the total amount of RNA polymerase in wild type cells at fast division rate is given by about 10 μM of which only 1 μM is free, while the rest is bound to the endogenous DNA. Since in the simulations we did not account for endogenous DNA, we assumed that only 1 μM is available in total to the two exogenous modules. A similar reasoning was employed to set $Y_{tot} = 10$ μM. Specifically, in exponential growth, we have about 34 μM of total ribosomes' concentration, but only about 30% of this is free, resulting in about 10 μM concentration of ribosomes available to the exogenous modules (http://bionumbers.hms.harvard.edu).

Figure 7.3a illustrates that as the activator concentration A increases, there is no substantial perturbation on the free amount of RNA polymerase. However, because the resulting perturbation on the free amount of ribosomes (Figure 7.3b) is significant, the resulting decrease of P_1 is substantial (Figure 7.3d).

7.2 Stochastic effects: Design tradeoffs in systems with large gains

As we have seen in Chapter 6, a biomolecular system can be rendered insensitive to retroactivity by implementing a large input amplification gain in a negative feedback loop. However, relying on high gains, this type of design may have undesired effects in the presence of noise, as seen in a different context in Section 5.2. Here, we employ the Langevin equation introduced in Chapter 4 to analyze this problem. Here, we treat the Langevin equation as a regular ordinary differential equation with inputs, allowing us to apply the tools described in Chapter 3.

Consider a system, such as the transcriptional component of Figure 6.4, in which a protein X is produced, degraded, and is an input to a downstream system, such as a transcriptional component. Here, we assume that both the production and the degradation of protein X can be tuned through a gain G, something that can be realized through the designs illustrated in Chapter 6. Hence, the production rate of X is given by a time-varying function $Gk(t)$ while the degradation rate is given by $G\gamma$.

The system can be simply modeled by the chemical reactions

$$0 \underset{G\gamma}{\overset{G\,k(t)}{\rightleftharpoons}} \text{X}, \qquad \text{X} + \text{p} \underset{k_{\text{off}}}{\overset{k_{\text{on}}}{\rightleftharpoons}} \text{C},$$

in which we assume that the binding sites p are in total constant amount denoted p_{tot}, so that $p + C = p_{tot}$.

We have shown in Section 6.5 that increasing the gain G is beneficial for attenuating the effects of retroactivity on the upstream component applied by the connected downstream system. However, as shown in Figure 7.4, increasing the gain G impacts the frequency content of the noise in a single realization. In particular, as G increases, the realization shows perturbations (with respect to the mean value) with higher frequency content.

To study this problem, we employ the Langevin equation (Section 4.1). For our system, we obtain (assuming unit volume for simplifying the mathematical

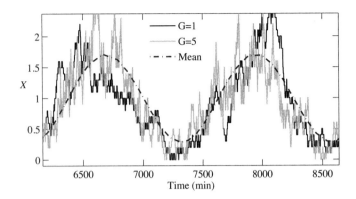

Figure 7.4: Stochastic simulations illustrating that increasing the value of G produces perturbations of higher frequency. Two realizations are shown with different values of G without load. The parameters used in the simulations are $\gamma = 0.01$ min^{-1} and the frequency of the input is $\omega = 0.005$ rad/min with input signal given by $k(t) = \gamma(1 + 0.8\sin(\omega t))$ nM/min. The mean of the signal is given as reference. Figure adapted from [49].

derivations):

$$\frac{dX}{dt} = Gk(t) - G\gamma X - k_{on}(p_{tot} - C)X + k_{off}C + \sqrt{Gk(t)}\,\Gamma_1(t) - \sqrt{G\gamma X}\,\Gamma_2(t)$$
$$- \sqrt{k_{on}(p_{tot} - C)X}\,\Gamma_3(t) + \sqrt{k_{off}C}\,\Gamma_4(t), \qquad (7.6)$$
$$\frac{dC}{dt} = k_{on}(p_{tot} - C)X - k_{off}C + \sqrt{k_{on}(p_{tot} - C)X}\,\Gamma_3(t) - \sqrt{k_{off}C}\,\Gamma_4(t).$$

The above system can be viewed as a nonlinear system with five inputs, $k(t)$ and $\Gamma_i(t)$ for $i = 1, 2, 3, 4$. Let $k(t) = \bar{k}$, and $\Gamma_1(t) = \Gamma_2(t) = \Gamma_3(t) = \Gamma_4(t) = 0$ be constant inputs and let $\bar{X}$ and $\bar{C}$ be the corresponding equilibrium points. Then for small amplitude signals $\tilde{k}(t) = k(t) - \bar{k}$ the linearization of the system (7.6) leads, with abuse of notation, to

$$\frac{dX}{dt} = G\tilde{k}(t) - G\gamma X - k_{on}(p_{tot} - \bar{C})X + k_{on}\bar{X}C + k_{off}C$$
$$+ \sqrt{G\bar{k}}\,\Gamma_1(t) - \sqrt{G\gamma \bar{X}}\,\Gamma_2(t) + \sqrt{k_{off}\bar{C}}\,\Gamma_4(t) - \sqrt{k_{on}(p_{tot} - \bar{C})\bar{X}}\,\Gamma_3(t),$$
$$\frac{dC}{dt} = k_{on}(p_{tot} - \bar{C})X - k_{on}\bar{X}C - k_{off}C - \sqrt{k_{off}\bar{C}}\,\Gamma_4(t) + \sqrt{k_{on}(p_{tot} - \bar{C})\bar{X}}\,\Gamma_3(t).$$

We can further simplify the above expressions by noting that $\gamma\bar{X} = \bar{k}$ and $k_{on}(p_{tot} - \bar{C})\bar{X} = k_{off}\bar{C}$. Also, since Γ_j are independent identical Gaussian white noise processes, we can write $\Gamma_1(t) - \Gamma_2(t) = \sqrt{2}N_1(t)$ and $\Gamma_3(t) - \Gamma_4(t) = \sqrt{2}N_2(t)$, in which $N_1(t)$ and $N_2(t)$ are independent Gaussian white noise processes identical to $\Gamma_j(t)$.

This simplification leads to the system

$$\frac{dX}{dt} = G\tilde{k}(t) - G\gamma X - k_{on}(p_{tot} - \bar{C})X + k_{on}\bar{X}C + k_{off}C + \sqrt{2G\bar{k}}N_1(t) - \sqrt{2k_{off}\bar{C}}N_2(t),$$

$$\frac{dC}{dt} = k_{on}(p_{tot} - \bar{C})X - k_{on}\bar{X}C - k_{off}C + \sqrt{2k_{off}\bar{C}}N_2(t). \tag{7.7}$$

This is a system with three inputs: the deterministic input $\tilde{k}(t)$ and two independent white noise sources $N_1(t)$ and $N_2(t)$. One can interpret N_1 as the source of the fluctuations caused by the production and degradation reactions while N_2 is the source of fluctuations caused by binding and unbinding reactions. Since the system is linear, we can analyze the different contributions of each noise source separately and independent from the signal $\tilde{k}(t)$.

We can simplify this system by taking advantage once more of the separation of time scales between protein production and degradation and the reversible binding reactions, defining a small parameter $\epsilon = \gamma/k_{off}$ and letting $K_d = k_{off}/k_{on}$. By applying singular perturbation theory, we can set $\epsilon = 0$ and obtain the reduced system on the slow time scale as performed in Section 6.3. In this system, the transfer function from N_1 to X is given by

$$H_{XN_1}(s) = \frac{\sqrt{2G\bar{k}}}{s(1+\bar{R}) + G\gamma}, \qquad \bar{R} = \frac{p_{tot}/K_d}{((\bar{k}/\gamma)/K_d + 1)^2}. \tag{7.8}$$

The zero frequency gain of this transfer function is equal to

$$H_{XN_1}(0) = \frac{\sqrt{2\bar{k}}}{\sqrt{G}\gamma}.$$

Thus, as G increases, the zero frequency gain decreases. But for large enough frequencies ω, $j\omega(1+\bar{R}) + G\gamma \approx j\omega(1+\bar{R})$, and the amplitude is approximately given by

$$|H_{XN_1}(j\omega)| \approx \frac{\sqrt{2\bar{k}G}}{\omega(1+\bar{R})},$$

which is a monotonically increasing function of G. This effect is illustrated in Figure 7.5. The frequency at which the amplitude of $|H_{XN_1}(j\omega)|$ computed with $G = 1$ intersects the amplitude $|H_{XN_1}(j\omega)|$ computed with $G > 1$ is given by the expression

$$\omega_e = \frac{\gamma\sqrt{G}}{(1+\bar{R})}.$$

Thus, when increasing the gain from 1 to $G > 1$, we reduce the noise at frequencies lower than ω_e but we increase it at frequencies larger than ω_e.

While retroactivity contributes to filtering noise in the upstream system as it decreases the bandwidth of the noise transfer function (expression (7.8)), high gains

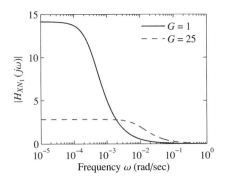

Figure 7.5: Magnitude of the transfer function $H_{XN_1}(s)$ as a function of the input frequency ω. The parameters used in this plot are $\gamma = 0.01$ min^{-1}, $K_d = 1$ nM, $k_{off} = 50$ min^{-1}, $\omega = 0.005$ rad/min, $p_{tot} = 100$ nM. When G increases, the contribution from N_1 decreases at low frequency but it spreads to a higher range of the frequency.

contribute to increasing noise at frequencies higher than ω_e. In particular, when increasing the gain from 1 to $G > 1$ we reduce the noise in the frequency ranges below ω_e, but the noise at frequencies above ω_e increases. If we were able to indefinitely increase G, we could send G to infinity attenuating the deterministic effects of retroactivity while amplifying noise only at very high, hence not relevant, frequencies.

In practice, however, the value of G is limited. For example, in the insulation device based on phosphorylation, G is limited by the amounts of substrate and phosphatase that we can have in the system. Hence, a design tradeoff needs to be considered when designing insulation devices: placing the largest possible G attenuates retroactivity but it may increase noise in a possibly relevant frequency range.

In this chapter, we have presented some of the tradeoffs that need to be accounted for when designing biomolecular circuits in living cells and focused on the problem of competition for shared resources and on noise. Problems of resource sharing, noise, and retroactivity are encompassed in a more general problem faced when engineering biological circuits, which is referred to as "context-dependence." That is, the functionality of a module depends on its context. Context-dependence is due to a number of different factors. These include unknown regulatory interactions between the module and its surrounding systems; various effects that the module has on the cell network, such as metabolic burden [12] and effects on cell growth [85]; and the dependence of the module's parameters on the specific biophysical properties of the cell and its environment, including temperature and the presence of nutrients. Future biological circuit design techniques will have to address all these additional problems in order to ensure that circuits perform robustly once interacting in the cellular environment.

Exercises

7.1 Assume that both Module 1 and Module 2 considered in Section 7.1 can be activated. Extend the analytical derivations of the text to this case.

7.2 A similar derivation to what was performed in Section 7.1 can be carried if R were a repressor of Module 2. Using a one-step reaction model for gene expression, write down the reaction equations for this case and the ordinary differential equations describing the rate of change of P_1 and P_2. Then, determine how the free concentration of RNA polymerase is affected by changes in R and how P_1 is affected by changes in R.

7.3 Consider again the case of a repressor as considered in the previous exercise. Now, consider a two-step reaction model for transcription and build a simulation model with parameter values as indicated in the text and determine the extent of coupling between Module 1 and Module 2 when the repressor is increased.

7.4 Consider the system (7.7) and calculate the transfer function from the noise source N_2 to X.

7.5 Consider the insulation device based on phosphorylation illustrated in Section 6.5. Perform stochastic simulations to investigate the tradeoff between retroactivity attenuation and noise amplification when key parameters are changed. In particular, you can perform one study in which the time scale of the cycle changes and a different study in which the total amounts of substrate and phosphatase are changed.

Bibliography

[1] K. J. Åström and R. M. Murray. *Feedback Systems: An Introduction for Scientists and Engineers.* Princeton University Press, 2008. Available at http://www.cds.caltech.edu/~murray/amwiki.

[2] B. Alberts, D. Bray, J. Lewis, M. Raff, K. Roberts, and J. D. Watson. *The Molecular Biology of the Cell.* Garland Science, 5th edition, 2008.

[3] R. Algar, T. Ellis, and G.-B. Stan. Modelling the burden caused by gene expression: an in silico investigation into the interactions between synthetic gene circuits and their chassis cell. *arXiv:1309.7798,* 2013.

[4] U. Alon. *An Introduction to Systems Biology. Design Principles of Biological Circuits.* Chapman-Hall, 2007.

[5] W. Arber and S. Linn. DNA modification and restriction. *Annual Review of Biochemistry,* 38:467–500, 1969.

[6] M. R. Atkinson, M. A. Savageau, J. T. Meyers, and A. J. Ninfa. Development of genetic circuitry exhibiting toggle switch or oscillatory behavior in *Escherichia coli. Cell,* pages 597–607, 2003.

[7] D. W. Austin, M. S. Allen, J. M. McCollum, R. D. Dar, J. R. Wilgus, G. S. Sayler, N. F. Samatova, C. D. Cox, and M. L. Simpson. Gene network shaping of inherent noise spectra. *Nature,* 2076:608–611, 2006.

[8] D. Baker, G. Church, J. Collins, D. Endy, J. Jacobson, J. Keasling, P. Modrich, C. Smolke, and R. Weiss. ENGINEERING LIFE: Building a FAB for biology. *Scientific American,* June 2006.

[9] N. Barkai and S. Leibler. Robustness in simple biochemical networks. *Nature,* 387(6636):913–917, 1997.

[10] A. Becskei and L. Serrano. Engineering stability in gene networks by autoregulation. *Nature,* 405:590–593, 2000.

[11] D. Bell-Pedersen, V. M. Cassone, D. J. Earnest, S. S. Golden, P. E. Hardin, T. L. Thomas, and M. J. Zoran. Circadian rhythms from multiple oscillators: lessons from diverse organisms. *Nature Reviews Genetics,* 6(7):544, 2005.

[12] W. E. Bentley, N. Mirjalili, D. C. Andersen, R. H. Davis, and D. S. Kompala. Plasmid-encoded protein: the principal factor in the "metabolic burden" associated with recombinant bacteria. *Biotechnol. Bioeng.,* 35(7):668–681, 1990.

[13] BioNumbers: The database of useful biological numbers. http://bionumbers.org, 2012.

[14] L. Bleris, Z. Xie, D. Glass, A. Adadey, E. Sontag, and Y. Benenson. Synthetic incoherent feedforward circuits show adaptation to the amount of their genetic template.
Molecular Systems Biology, 7:519, 2011.

[15] H. Bremer and P. Dennis. Modulation of chemical composition and other parameters of the cell by growth rate. In *Escherichia coli and Salmonella: Cellular and
Molecular Biology* (edited by F. C. Neidhart et al.), ASM Press, Washington, DC,
183:1553–1569, 1996.

[16] H. Bremer, P. P. Dennis, and M. Ehrenberg. Free RNA polymerase and modeling
global transcription in *Escherichia coli. Biochimie*, 85(6):597–609, 2003.

[17] C. I. Byrnes, F. D. Priscoli, and A. Isidori. *Output Regulation of Uncertain Nonlinear Systems*. Birkhauser, 1997.

[18] B. Canton, A. Labno, and D. Endy. Refinement and standardization of synthetic
biological parts and devices. *Nature Biotechnology*, 26(7):787–793, 2008.

[19] M. Chalfie, Y. Tu, G. Euskirchen, W. Ward, and D. Prasher. Green fluorescent
protein as a marker for gene expression. *Science*, 263(5148):802–805, 1994.

[20] A. J. Courey. *Mechanisms in Transcriptional Regulation*. Wiley-Blackwell, 2008.

[21] D. Del Vecchio. Design and analysis of an activator-repressor clock in *e. coli*. In
Proc. American Control Conference, 2007.

[22] D. Del Vecchio, A. J. Ninfa, and E. D. Sontag. Modular cell biology: Retroactivity
and insulation. *Nature/EMBO Molecular Systems Biology*, 4:161, 2008.

[23] L. N. M. Duysens and J. Amesz. Fluorescence spectrophotometry of reduced
phosphopyridine nucleotide in intact cells in the near-ultraviolet and visible region.
Biochim. Biophys. Acta, 24:19–26, 1957.

[24] H. El-Samad, J. P. Goff, and M. Khammash. Calcium homeostasis and parturient
hypocalcemia: An integral feedback perspective. *J. Theoret. Biol.*, 214:17–29, 2002.

[25] S. P. Ellner and J. Guckenheimer. *Dynamic Models in Biology*. Princeton University
Press, Princeton, NJ, 2005.

[26] M. B. Elowitz and S. Leibler. A synthetic oscillatory network of transcriptional
regulators. *Nature*, 403(6767):335–338, 2000.

[27] M. B. Elowitz, A. J. Levine, E. D. Siggia, and P. Swain. Stochastic gene expression
in a single cell. *Science*, 297(5584):1183–1186, 2002.

[28] D. Endy. Foundations for engineering biology. *Nature*, 438:449–452, 2005.

[29] K. M. Eveker, D. L. Gysling, C. N. Nett, and O. P. Sharma. Integrated control of
rotating stall and surge in high-speed multistage compression systems. *Journal of
Turbomachinery*, 120(3):440–445, 1998.

[30] T. S. Gardner, C .R. Cantor, and J. J. Collins. Construction of the genetic toggle
switch in *Escherichia coli. Nature*, 403:339–342, 2000.

[31] D. C. Gibson, J. I. Glass, C. Lartigue, V. N. Noskov, R. Y. Chuang, M. A. Algire, G. A. Benders, M. G. Montague, L. Ma, M. M. Moodie, C. Merryman, S. Vashee, R. Krishnakumar, N. Assad-Garcia, C. Andrews-Pfannkoch, E. A. Denisova, L. Young, Z.-Q. Qi, T. H. Segall-Shapiro, C. H. Calvey, P. P. Parmar, C. A. Hutchison, H. O. Smith, and J. C. Venter. Creation of a bacterial cell controlled by a chemically synthesized genome. *Science*, 329(5987):52–56, 2010.

[32] D. T. Gillespie. *Markov Processes: An Introduction For Physical Scientists*. Academic Press, 1976.

[33] D. T. Gillespie. Exact stochastic simulation of coupled chemical reactions. *Journal of Physical Chemistry*, 81(25):2340–2361, 1977.

[34] D. T. Gillespie. The chemical Langevin equation. *Journal of Chemical Physics*, 113(1):297–306, 2000.

[35] L. Goentoro, O. Shoval, M. W. Kirschner, and U. Alon. The incoherent feedforward loop can provide fold-change detection in gene regulation. *Molecular Cell*, 36:894–899, 2009.

[36] A. Goldbeter and D. E. Koshland. An amplified sensitivity arising from covalent modification in biological systems. *Proc. of the National Academy of Sciences*, pages 6840–6844, 1981.

[37] J. Greenblatt, J. R. Nodwell, and S. W. Mason. Transcriptional antitermination. *Nature*, 364(6436):401–406, 1993.

[38] I. L. Grigiriva, N. J. Phleger, V. K. Mutalik, and C. A. Gross. Insights into transcriptional regulation and σ competition from an equilibrium model of RNA polymerase binding to DNA. *Proc. of the National Academy of Sciences*, 103(14):5332–5337, 2006.

[39] J. Guckenheimer and P. Holmes. *Nonlinear Oscillations, Dynamical Systems, and Bifurcations of Vector Fields*. Springer, 1983.

[40] S. Hastings, J. Tyson, and D. Webster. Existence of periodic solutions for negative feedback cellular systems. *J. of Differential Equations*, 25:39–64, 1977.

[41] R. Heinrich, B. G. Neel, and T. A. Rapoport. Mathematical models of protein kinase signal transduction. *Molecular Cell*, 9:957–970, 2002.

[42] B. Hess, A. Boiteux, and J. Kruger. Cooperation of glycolytic enzymes. *Adv. Enzyme Regul*, 7:149–167, 1969.

[43] A. Hilfinger and J. Paulsson. Separating intrinsic from extrinsic fluctuations in dynamic biological systems. *Proc. of the National Academy of Sciences*, 108(29):12167–12172, 2011.

[44] C. F. Huang and J. E. Ferrell. Ultrasensitivity in the mitogen-activated proteinkinase cascade. *Proc. of the National Academy of Sciences*, 93(19):10078–10083, 1996.

[45] T. P. Hughes. *Elmer Sperry: Inventor and Engineer*. Johns Hopkins University Press, Baltimore, MD, 1993.

[46] R. S. Cox III, M. G. Surette, and M. B. Elowitz. Programming gene expression with combinatorial promoters. *Mol. Syst. Biol.*, 3:145, 2007.

[47] A. Ishihama. Functional modulation of *E. coli* RNA polymerase. *Ann. Rev. Microbiol.*, 54:499–518, 2000.

[48] F. Jacob and J. Monod. Genetic regulatory mechanisms in the synthesis of proteins. *J. Mol. Biol.*, 3:318–356, 1961.

[49] S. Jayanthi and D. Del Vecchio. On the compromise between retroactivity attenuation and noise amplification in gene regulatory networks. In *Proc. Conference on Decision and Control*, pages 4565–4571, 2009.

[50] S. Jayanthi and D. Del Vecchio. Retroactivity attenuation in bio-molecular systems based on timescale separation. *IEEE Transactions on Automatic Control*, 56:748–761, 2011.

[51] S. Jayanthi, K. Nilgiriwala, and D. Del Vecchio. Retroactivity controls the temporal dynamics of gene transcription. *ACS Synthetic Biology*, DOI: 10.1021/sb300098w, 2013.

[52] P. Jiang, A. C. Ventura, S. D. Merajver, E. D. Sontag, A. J. Ninfa, and D. Del Vecchio. Load-induced modulation of signal transduction networks. *Science Signaling*, 4(194):ra67, 2011.

[53] N. G. Van Kampen. *Stochastic Processes in Physics and Chemistry*. Elsevier, 1992.

[54] A. S. Khalil and J. J. Collins. Synthetic biology: Applications come of age. *Nature Reviews Genetics*, 11(5):367, 2010.

[55] H. K. Khalil. *Nonlinear Systems*. Macmillan, 1992.

[56] E. Klipp, W. Liebermeister, C. Wierling, A. Kowald, H. Lehrach, and R. Herwig. *Systems Biology: A Textbook*. Wiley-VCH, 2009.

[57] S. Klumpp and T. Hwa. Growth-rate-dependent partitioning of RNA polymerases in bacteria. *Proc. of the National Academy of Sciences*, 105(51):20245–20250, 2008.

[58] P. Kundur. *Power System Stability and Control*. McGraw-Hill, New York, 1993.

[59] M. T. Laub, L. Shapiro, and H. H. McAdams. Systems biology of *Caulobacter*. *Annual Review of Genetics*, 51:429–441, 2007.

[60] J.-C. Leloup and A. Goldbeter. A molecular explanation for the long-term supression of circadian rhythms by a single light pulse. *American Journal of Physiology*, 280:1206–1212, 2001.

[61] J. J. Lemke, P. Sanchez-Vazquez, H. L. Burgos, G. Hedberg, W. Ross, and R. L. Gourse. Direct regulation of *Escherichia coli* ribosomal protein promoters by the transcription factors ppGpp and DksA. *Proc. of the National Academy of Sciences*, pages 1–6, 2012.

[62] W. Lohmiller and J. J. E. Slotine. On contraction analysis for non-linear systems. *Automatica*, 34:683–696, 1998.

[63] H. Madhani. *From a to alpha: Yeast as a Model for Cellular Differentiation*. CSHL Press, 2007.

[64] J. Mallet-Paret and H. L. Smith. The Poincaré-Bendixson theorem for monotone cyclic feedback systems. *J. of Differential Equations*, 2:367–421, 1990.

[65] J. E. Marsden and M. J. Hoffman. *Elementary Classical Analysis*. Freeman, 2000.

[66] J. E. Marsden and M. McCracken. *The Hopf Bifurcation and Its Applications*. Springer-Verlag, New York, 1976.

[67] S. Marsigliante, M. G. Elia, B. Di Jeso, S. Greco, A. Muscella, and C. Storelli. Increase of [Ca(2+)](i) via activation of ATP receptors in PC-Cl3 rat thyroid cell line. *Cell. Signal*, 14:61–67, 2002.

[68] H. H. McAdams and A. Arkin. Stochastic mechanisms in gene expression. *Proc. of the National Academy of Sciences*, 94:814–819, 1997.

[69] C. R. McClung. Plant circadian rhythms. *Plant Cell*, 18:792–803, 2006.

[70] M. W. McFarland, editor. *The Papers of Wilbur and Orville Wright*. McGraw-Hill, New York, 1953.

[71] P. Miller and X. J. Wang. Inhibitory control by an integral feedback signal in prefrontal cortex: A model of discrimination between sequential stimuli. *Proc. of the National Academy of Sciences*, 103:201–206, 2006.

[72] C. J. Morton-Firth, T. S. Shimizu, and D. Bray. A free-energy-based stochastic simulation of the tar receptor complex. *Journal of Molecular Biology*, 286(4):1059–1074, 1999.

[73] J. D. Murray. *Mathematical Biology*, Vols. I and II. Springer-Verlag, New York, 3rd edition, 2004.

[74] C. J. Myers. *Engineering Genetic Circuits*. Chapman and Hall/CRC Press, 2009.

[75] T. Nagashima, H. Shimodaira, K. Ide, T. Nakakuki, Y. Tani, K. Takahashi, N. Yumoto, and M. Hatakeyama. Quantitative transcriptional control of ErbB receptor signaling undergoes graded to biphasic response for cell differentiation. *J. Biol. Chem.*, 282:4045–4056, 2007.

[76] R. Nesher and E. Cerasi. Modeling phasic insulin release: Immediate and time-dependent effects of glucose. *Diabetes*, 51:53–59, 2002.

[77] H. R. Ossareh, A. C. Ventura, S. D. Merajver, and D. Del Vecchio. Long signaling cascades tend to attenuate retroactivity. *Biophysical Journal*, 10:1617–1626, 2011.

[78] R. Phillips, J. Kondev, and J. Theriot. *Physical Biology of the Cell*. Garland Science, 2009.

[79] M. Ptashne. *A Genetic Switch*. Blackwell Science, Inc., 1992.

[80] P. E. M. Purnick and R. Weiss. The second wave of synthetic biology: From modules to systems. *Nature Reviews Molecular Cell Biology*, 10(6):410–422, 2009.

[81] E. K. Pye. Periodicities in intermediary metabolism. *Biochronometry*, pages 623–636, 1971.

[82] L. Qiao, R. B. Nachbar, I. G. Kevrekidis, and S. Y. Shvartsman. Bistability and oscillations in the Huang-Ferrell model of MAPK signaling. *PLoS Computational Biology*, 3:e184, 2007.

[83] C. V. Rao, J. R. Kirby, and A. P. Arkin. Design and diversity in bacterial chemotaxis: A comparative study in *Escherichia coli* and *Bacillus subtilis*. *PLoS Biology*, 2(2):239–252, 2004.

[84] H. M. Sauro and B. N. Kholodenko. Quantitative analysis of signaling networks. *Progress in Biophysics & Molecular Biology*, 86:5–43, 2004.

[85] M. Scott, C. W. Gunderson, E. M. Mateescu, Z. Zhang, and T. Hwa. Interdependence of cell growth and gene expression: Origins and consequences. *Science*, 330:1099–1202, 2010.

[86] D. E. Seborg, T. F. Edgar, and D. A. Mellichamp. *Process Dynamics and Control*. Wiley, Hoboken, NJ, 2nd edition, 2004.

[87] Thomas S Shimizu, Yuhai Tu, and Howard C Berg. A modular gradient-sensing network for chemotaxis in *Escherichia coli* revealed by responses to time-varying stimuli. *Molecular Systems Biology*, 6:382, 2010.

[88] O. Shimomura, F. Johnson, and Y. Saiga. Extraction, purification and properties of Aequorin, a bioluminescent protein from the luminous hydromedusan, *Aequorea. J. Cell. Comp. Physiol.*, 59(3):223–239, 1962.

[89] O. Shoval, U. Alon, and E. Sontag. Symmetry invariance for adapting biological systems. *SIAM J. Applied Dynamical Systems*, 10:857886, 2011.

[90] E. D. Sontag. *Mathematical Control Theory: Deterministic Finite Dimensional Systems*. Springer, New York, 2nd edition, 1998.

[91] E. D. Sontag. Remarks on feedforward circuits, adaptation, and pulse memory. *IET Systems Biology*, 4:39–51, 2010.

[92] P. S. Swain, M. B. Elowitz, and E. D. Siggia. Intrinsic and extrinsic contributions to stochasticity in gene expression. *Proc. of the National Academy of Sciences*, 99(20):12795–12800, 2002.

[93] J. Tsang, J. Zhu, and A. van Oudenaarden. Microrna-mediated feedback and feedforward loops are recurrent network motifs in mammals. *Mol. Cell.*, 26:753–767, 2007.

[94] K. V. Venkatesh, S. Bhartiya, and A. Ruhela. Mulitple feedback loops are key to a robust dynamic performance of tryptophan regulation in *Escherichia coli*. *FEBS Letters*, 563:234–240, 2004.

[95] A. C. Ventura, P. Jiang, L. van Wassenhove, D. Del Vecchio, S. D. Merajver, and A. J. Ninfa. The signaling properties of a covalent modification cycle are altered by a downstream target. *Proc. of the National Academy of Sciences*, 107(22):10032–10037, 2010.

[96] O. S. Venturelli, H. El-Samad, and R. M. Murray. Synergistic dual positive feedback loops established by molecular sequestration generate robust bimodal response. *Proc. of the National Academy of Sciences*, 109(48):E3324–3333, 2012.

[97] L. Villa-Komaroff, A. Efstratiadis, S. Broome, P. Lomedico, R. Tizard, S. P. Naber, W. L. Chick, and W. Gilbert. A bacterial clone synthesizing proinsulin. *Proc. of the National Academy of Sciences*, 75(8):3727–3731, 1978.

[98] J. Vind, M. A. Sørensen, M. D. Rasmussen, and S. Pedersen. Synthesis of proteins in *Escherichia coli* is limited by the concentration of free ribosomes. *J. Mol. Biol.*, 231:678–688, 1993.

[99] C. A. Voigt. Genetic parts to program bacteria. *Current Opinions in Biotechnology*, 17(5):548–557, 2006.

[100] S. Wiggins. *Introduction to Applied Nonlinear Dynamical Systems and Chaos.* Springer, 2003.

[101] L. Yang and P. A. Iglesias. Positive feedback may cause the biphasic response observed in the chemoattractant-induced response of Dictyostelium cells. *Systems Control Lett.*, 55:329–337, 2006.

[102] T.-M. Yi, Y. Huang, M. I. Simon, and J. C. Doyle. Robust perfect adaptation in bacterial chemotaxis through integral feedback control. *Proc. of the National Academy of Sciences*, 97(9):4649–4653, 2000.

[103] N. Yildirim and M. C. Mackey. Feedback regulation in the lactose operon: A mathematical modeling study and comparison with experimental data. *Biophysical Journal*, 84(5):2841–2851, 2003.

Index